HOW TO Restore CAMARO 1967-1969

Tony E. Huntimer & Brian Henderson

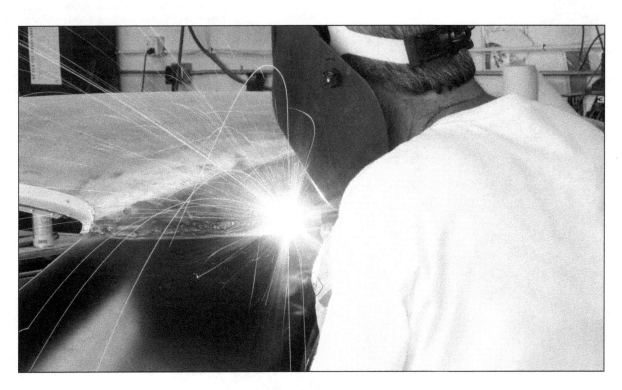

CarTech®

CarTech®
CarTech®, Inc.
39966 Grand Avenue
North Branch, MN 55056
Phone: 651-277-1200 or 800-551-4754
Fax: 651-277-1203
www.cartechbooks.com

© 2010 by Tony E. Huntimer & Brian Henderson

All rights reserved. No part of this publication may be reproduced or utilized in any form or by any means, electronic or mechanical, including photocopying, recording, or by any information storage and retrieval system, without prior permission from the Publisher. All text, photographs, and artwork are the property of the Author unless otherwise noted or credited.

The information in this work is true and complete to the best of our knowledge. However, all information is presented without any guarantee on the part of the Author or Publisher, who also disclaim any liability incurred in connection with the use of the information and any implied warranties of merchantability or fitness for a particular purpose. Readers are responsible for taking suitable and appropriate safety measures when performing any of the operations or activities described in this work.

All trademarks, trade names, model names and numbers, and other product designations referred to herein are the property of their respective owners and are used solely for identification purposes. This work is a publication of CarTech, Inc., and has not been licensed, approved, sponsored, or endorsed by any other person or entity. The Publisher is not associated with any product, service, or vendor mentioned in this book, and does not endorse the products or services of any vendor mentioned in this book.

Edit by Paul Johnson
Layout by Chris Fayers

ISBN 978-1-61325-224-6
Item No. SA178P

Library of Congress Cataloging-in-Publication Data

Huntimer, Tony E.
  How to restore your Camaro, 1967-1969 / by Tony E. Huntimer and Brian Henderson.
      p. cm.
  ISBN 978-1-934709-10-8
  1. Camaro automobile–Conservation and restoration. 2. Antique and classic cars–Conservation and restoration. I. Henderson, Brian, 1960- II. Title.
  TL215.C33H86 2010
  629.28'722–dc22

2009050382

Written, edited, and designed in the U.S.A.
Printed in U.S.A.

Title page:
With everything tacked in place, Dick was ready to start welding the quarter and the roof panel together. The weld between the quarter panel and the roof seam must be very strong and sound because it is a high-stress area and you don't want the seam to flex. Be very careful not to overheat the roof or the quarter because you can easily warp them, which adds even more time to your bodywork to fix. Dick welds a short section and then cools it with an air nozzle as he goes. (Photo courtesy Steven Rupp)

Back Cover Photos

Top Left:
Overall this quarter panel was in pretty good shape, but it was installed about 1/4 inch too low and needed to be corrected. However, a decent job filling the gaps with body filler was performed. Dick Kvamme of Best of Show Coachworks (BOS) removes the bulk of the panel with a plasma cutter. He left the roof seam, door jamb, and edges to be removed more carefully and trim the rest with a cutoff wheel after cutting the spot welds loose. (Photo courtesy Steven Rupp)

Top Right:
Applying chemical stripper is one of the methods of removing paint and filler from the body. Make sure you completely remove all the chemicals from the nooks, crannies, and cracks and neutralize the chemicals before applying primer or you'll have a huge mess. (Photo courtesy Detroit Speed)

Middle Left:
You need to soak your wet-or-dry paper in a bucket of water with a little ivory soap mixed in before sanding. Best of Show Coach Works employee, Jon Lindstrom, has started sanding with 600-grit paper wrapped around a rubber sanding pad. If you have five thick coats of clear over the stripes, you may have enough paint to sand the clear down to where the ridges on the borders are gone. In that case, you should start with a more aggressive 400-grit on the stripe edge and quickly move to 600-grit. (Photo courtesy Steven Rupp)

Middle Right:
If you're working on a car with a limited-slip differential, loosely install the caps with bolts that are approximately 1.5 inches longer than stock. With the caps loosely installed, pry the carrier loose and pop it out of the housing but not onto the floor or you may chip a tooth off the gear. If the differential is out of the car, it's easier to work on it without worrying about the carrier falling on the floor.

Bottom Left:
Starting in 1968, the ever-popular Houndstooth seat inserts were offered with the Deluxe interior. In 1968 you could only get Houndstooth with black or pearl parchment vinyl borders, but in 1969 you could order additional borders and Houndstooth colors.

Bottom Right:
A professionally restored dashboard area looks like this. Earlier in 1969, a walnut wheel was an option, which was replaced with this Rosewood steering wheel about halfway through the model year. This is an SS/RS 396 with the 120-mph speedo, 7,000-rpm tach with 6,000 rpm redline, and clock. (Photo courtesy Tony Lucas Car Collection)

CONTENTS

Acknowledgments 5
Introduction 6

Chapter 1: Getting to Know the Camaro 7
Valuable Camaros 8
Acquiring a Camaro 10
Identifying a Fake 16
Identifying Your VIN Plate 17
Deciphering the Cowl Tag 18

Chapter 2: Getting Started 20
Skills 20
Tools and Tool Basics 21
Materials and Specialty Tools 21
Where Do I Start? 21
Repair or Replace? 22
Game Plan 22
The Process 23
Original or Reproduction? 24
Parts Sources 25
Documentation 25
Organization 26
Disassembly 26

Chapter 3: Bodywork 28
What Should Be Replaced? 28
Common Rust Areas 29
Floorpan Replacement 31
Sound-Deadener Removal 32
Quarter Panels 32
Body Panel Alignment 36
Dings and Dents 37
Paint Removal 38
Body Stamp Care 40
Cowl Tag Removal 40

Chapter 4: The Painting Process 41
Can You Paint? 41
Why are Shops so Expensive? 42
What Can Go Wrong? 42
Decisions about the Outcome 43
How Correct? 43
Interior Paint 44
Underbody Paint 45
Firewall Paint 45
Cowl Paint Differences 46
Black Cowl Paint to Firewall Transition 47
Trunk Paint 47
Spraying Format 48
Spraying Disassembled Parts 49
Masking 49
Steps Before Spraying 49
Black-Out Paint 50
Stripes 50
Vinyl Top 54
Powder Coating 54
Scuff and Buff 55

Chapter 5: Engine Rebuilding 58
Selecting a Machine Shop 59
Identification 59
Inspecting and Preparing Parts 61
Engine-Rebuilding Lubricants 64
Dampers 71
Gaskets 72
Picking Parts 72
Extrude Honing 74
Engine Fasteners 74
Factory Engine Detailing 74
Factory Clamps 74

Chapter 6: Transmissions 75
Removal Safety 75
Manual Transmissions 76
4-Speed Manufacturers 77
Overdrive Transmissions 87
Automatic Transmissions 87

Chapter 7: Differentials 88
Inspection 88
Aftermarket Gear Sets 89
Gear Swap and Rebuild 89
Backlash 95
Test Drive 95
Identifying Parts 96
Identifying Gear Ratio 96

Chapter 8: Brakes 97
  Original Components and Restoration 97
  Brake Hardlines 98
  Master Cylinder 98
  Hold-Off Valve 99
  Brake Distribution Block (with brake warning) 99
  Pressure Regulator Valve 100
  OEM Brakes 101
  Home-Brewed 4-Wheel Disc Brakes 102
  Aftermarket Components 102
  Install the Master Cylinder 108
  Proportioning Valve 109
  Residual Pressure Valve 109

Chapter 9: Front Suspension 111
  Pre-Disassembly Inspection 111
  Separation Anxiety 112
  Front Suspension Tools 114
  Front Suspension Removal 114
  Subframe Changes 115
  Subframe and Control Arm Problems 116
  Painting Suspension Parts 117
  Rebuilding the Subframe 118
  Replacing Body Bushings 119
  Installing the Subframe 119

Chapter 10: Rear Suspension 122
  1967 122
  1968 123
  1969 123
  Mono-Leaf vs. Multi-Leaf 123
  Spiral Shocks 123
  Aftermarket Shocks 123
  Spring Perches 124
  Leaf Spring Removal 125
  Isolators In or Out 126
  Leaf Spring Bushing Types 126
  Performance Leaf Springs 127
  Installing Leaf Springs 127
  Rear Sway Bar 128

Chapter 11: Electrical 135
  Safety Tips 135
  Battery 136
  Installing the Harness 136
  Ignition 138
  Ignition Coils 139
  Tools 139
  Non-Factory Wiring 141
  Relay 141
  Electrical Components 141
  Aftermarket Accessories 142

Chapter 12: Interior 143
  Headliner 143
  Kick Panels 147
  Interior Paint 147
  Door Panels 147
  Door Mechanicals 149
  Rebuiding a Window Mechanism 150
  Water Shields 151
  Water Drains 152
  Carpet 153
  Dash Pad 154
  Console 154
  Seats 155
  Headrests 156
  Steering Wheel 156
  Seat Belts 157
  Sill Plates 158
  Repairing Plastic Parts 159

Chapter 13: Miscellaneous Mechanicals and Other Items 160
  Cooling System 160
  Cowl Hood 164
  Fresh-Air System 164
  Air Conditioning 165
  Carburetor 165
  Fuel Line Sizes 166
  Fuel Pump Installation 166
  Hood Latches and Hinges 167
  Endura Bumper 168
  Vinyl Top 169
  Convertible Top 169
  Wheels 170
  Tires 173
  Water Control 174
  Rubber Bumpers 175

Source Guide 176

ACKNOWLEDGMENTS

I couldn't have written this book without the dedication and assistance of world-class restorer Brian Henderson. His expansive knowledge of first-generation Camaros and factory restoration details was invaluable. He and I thank his wife, Pam, and his son, David, and his daughter, Lauren, for their support through the days and nights of pressing through restorations and assisting with this book. Brian also gives thanks to other individuals that I have also thanked, so they are included below.

I thank all of the people who have helped in one way or another or influenced my life in some manor to give me the ability to pass this information onto others. I hope I didn't leave anyone out and that I spelled their names correctly. To start, I thank my wife, Vikki Bristol-Huntimer, for giving me the support I needed to write this book. In no specific order, I also thank Robert and Maureen Cera, Slim, Yetta, Ken, Linda, and the entire Huntimer family, Steve and Martha Sanford, Cliff S. Witham, Jim Witham, Paul Caselas, Tracy Edmonds, Jeff Harwell, Chris Mead, Ed Matthews, Chris Fogarty, Brain Henderson, Joe Swezey, Frank Arone, Tony Lucas, Yenko.net, Supercar Registry, Bob Jones, Tom and Rob Clary, David Boland, Anthony Wilson, Sanders Wilhelm, Tom Zimmerman, Mary and David Pozzi, Don Roy, Camaros.net, Dan Kahn, Ken Sink, Cam Douglass, Jimi Day, Kristin Rogers, Becky Pias, Mark Schwartz, Doug Bracey, Dennis Dunio, Rich Deans, Vic DeLeon Sr and Jr., Ben Lingard, Ben Chase, Dan Schnider, Keith Spain, Paul Johnson, Ken Lucas, Eric Pogue, Jeff Smith, Steven Rupp, Nick Licata, Patrick Chaves, Keith Morita, John Hotchkis, Robert Cancilla, Chris Raschke, Wade Caldwell, Trish Yunick-Brown, Mark Dodge, Dave Stroncek, Charley Lillard, Ann Skrycki-Mohler, Ron Rotunno, David Logan, Mark Vogt, Billy Morris, Forrest Macomber, Joe Richardson, Robert Stewart, Mike Stasko, Steve Pardini, Mike Pomeroy, chevroletcamaros.com, Ed Bertrand, Larry Brogdin, Lake C. Speed Jr., Jeff Theobald, Rick Love, Bob Gromm, Dave Wiley, Jeff Chlupsa, John Lipori, Carl Casanova, Jay Rowlands, Larry Callahan, Marilyn Patton, Mrs. Reid, Dave Monyhan, Randy Oldham, David Gravley, Mike McGee, Bill DeFer, Kevin Stearns, Eric Detaeye, Jim Krolack, Karl Chicca, Mark Deshetler, Mark Zanella, Robert Cancilla, Cliff Burton, James Hetfield, Lars Ulrich, Kirk Hammett, Robert Trujillo, Ben O'Connor, Bill Koppinger, Wayne Mendoza, Bob Taggart, Kyle and Stacy Tucker, Gary Bohanick, and Adam Carolla.

It would have been tough to write this book without the help and support from the people who own and are members of websites that support the hobby of rebuilding, restoring, and modifying Camaros. These include Yenko.net, Camaros.net, Pro-Touring.com, and Lateral-G.net. These four websites were willing to participate and felt it would benefit other hobbyists, and allowed us to utilize data and information during the process of our research.

Thanks to everyone who helped.

INTRODUCTION

Many so-called restoration guides have been published over the past 30 years, but, in many instances, these "restoration guides" have only provided authenticity information, such as identifying the correct engine color code or badges for a particular model and year. This type of information certainly has value, since factory-correct restored cars are more appealing and obviously more valuable; however, these books often lack the information and instruction for actually restoring an entire car.

My goal with How to Restore Your Camaro 1967-1969 is to provide the proper procedures and actual nuts-and-bolts instruction for restoring a first-generation 1967–1969 Camaro to like-new condition. As such, this is a real-world guide for restoring a collector car that will be driven and enjoyed, but just to be clear, this is not a guide for a 100-point concours restoration on a Camaro. A restoration of that scope and expense is well beyond many enthusiasts' budget and ability level, and, therefore, is not relevant for this book. With that stated, this book does contain a sprinkling of concours restoration details and tips to raise the bar of your restoration if you choose to use them. Some of these details have never been written about in a book and have been closely guarded by top restorers to keep a leg up on the competition.

I have conducted extensive research to show you the most useful techniques for identifying a Camaro and verifying its authenticity. Few things are more heartbreaking than buying a collector car that was misrepresented and worth far less than what you paid for it. No one likes to be had. I have invested considerable time discussing the tools, materials, and facilities required for different stages of the restoration. And, as always, planning is a critical stage of any restoration. It allows you to make the best use of your time and attain the best results.

Restoring an entire car is not for the casual mechanic or inattentive enthusiast. If it were easy, everyone would be restoring their own car, and there wouldn't be professional restoration shops. While the intent isn't to scare you, you do need to determine your time, ability level, and budget. Simply put, it's hard to be an expert at everything. While one person may have experience and talent in engine building, he or she may not be an experienced body worker or painter. Therefore, very few people can restore an entire car from bumper to bumper all on their own. Hence, you need to figure out the restoration procedures you can do and make arrangements with a qualified shop to perform the work you will not be doing.

The end goal is to accomplish a professional-caliber restoration that you are proud of. In order to help you reach your target, this is the most informative and authoritative restoration guide possible in the available space. Co-author Brian Henderson and I have provided information that goes into exceptional depth and detail, but we did not have the space to go into exhaustive detail on every aspect of restoration. Therefore, we covered the most important aspects of engine and transmission rebuilding, but there are entire books written on these subjects, so we couldn't provide extremely comprehensive coverage for these components. However, I believe the perspective and real-world information is indispensable when performing a restoration. Each chapter provides major considerations when restoring a particular component group, the typical challenges and possible pitfalls, and the required tools and materials. But, most importantly, it gives you real hands-on information for a particular restoration task. Many of the important restoration procedures are presented in step-by-step format, so you can confidently and professionally perform the specified procedure.

In order to give you the most amount of information, Brian and I have covered restoration procedures on a variety of Camaros. This should give you the guidance and knowledge to complete a restoration. We cover engine and transmission rebuilding, bodywork, painting, drivetrain restoration, electrics, suspension, steering, interior, and much more. While restoring a Camaro yourself may seem like a Herculean task, the sense of satisfaction and accomplishment that comes with completing a faithful and professional restoration simply cannot be equaled.

And while the restorations do not come without challenges, it should be a fascinating and productive journey. A first-time restorer will acquire a vast amount of knowledge and skill that he or she didn't possess before the project began. You have taken on the admirable task of restoring an automotive icon—the Camaro. You are preserving a piece of automotive history, and, for that, I salute you.

CHAPTER 1

GETTING TO KNOW THE CAMARO

Because there are plenty of books available that cover just about every aspect of the history behind the Camaro, this chapter gives a brief overview of Camaro history before we move on to how to go about restoring one.

The first Camaro was introduced to the public on September 29, 1966, according to Mike Antonick, author of *Camaro White Book*. The Camaro was GM's response to Ford's Mustang, and this worthy competitor became an automotive icon of its own. Chevrolet made huge efforts in stepping outside the box when designing the Camaro. It was the first car to utilize a front subframe mounted to a unitized body with rubber bushings to isolate road noise and vibration. This gave the Camaro a more refined feel compared to other cars competing in the same market segment. The car was large enough to seat four people, but small enough with optional powerplants to be truly competitive with other manufacturers' cars on the street and the track.

These were the days of "race on Sunday and sell on Monday." To help boost sales, Chevrolet spent a lot of time, money, and energy covertly supporting the Camaro in SCCA racing. Entrepreneurs funded most of the drag-racing campaigns without help from the factory. Vince Piggins, Bill Howell, Paul King, Paul Prior, Jim Travers (TRACO), Smokey Yunick, Don Yenko, Fred Gibb, and many more were all in the right place at the right time to make the Camaro an American icon.

The Camaro had many names associated with it during initial design—Chaparral, Panther, and a slew of others until the name Camaro was announced in June 1966 at a press conference. This was extremely late in product development considering only three months later, the public was seeing a completed car for the first time.

Some changes were made between the 1967 and 1968 model years, and the untrained eye may not see the differences without seeing the

Even a dilapidated Z28 like this one could be worth more than a poorly restored one because all the parts are original equipment and the original finishes are probably still apparent under some of the dirt. (Photo courtesy of Dennis Doherty)

CHAPTER 1

Camaro Model Changes 1967–1969

Many details differed from 1967 to 1969, from the VIN (vehicle identification number) plate location change from 1967 to 1968 to the center body-mounting pad design change in 1968 on the front-subframe. Here are others:

1967 Features
- Non-staggered shock absorbers
- Cast-iron door hinges
- No side marker lights
- Electrically operated RS headlight doors

1968 Features
- Staggered rear shocks replaced the non-staggered shocks
- Front bumper mounting locations changed in an effort to increase effectiveness of the front bumper
- Stamped-steel door hinges replaced the cast-iron hinges of 1967
- Side marker lights were added
- Different grille and headlights
- Vacuum-operated RS headlight doors
- Different taillight housings
- Vent wing window of the 1967 was eliminatd
- Astro ventilation was added
- RPO D80 (front and rear spoilers) was added
- Interior side panels changed so top edge is covered with vinyl
- SS hood inserts changed to simulated injector stacks

1969 Features
- ZL2 cowl induction hood was introduced
- Endura front bumper
- Almost a complete sheetmetal and interior (dash, door panels, interior door lock location, etc.) makeover below the windows
- JL8 4-wheel disc brakes were added as an RPO

cars parked next to each other. Major design changes were made to the 1969 Camaro body and dashboard.

Even though first-generation Camaros are very valuable compared to most other 1967 through 1969 GM models, some Camaros are even more valuable than others.

Valuable Camaros

The most desirable first-generation Camaro options for a collector are: COPOs (of these Camaros, the most well known is the ZL1), Pace Cars, Vintage Trans-Am racing cars, Z28s, SSs, RSs, and JL8s.

COPO Camaros

COPO stands for Central Office Production Order. The ultimate big-block street Camaros and many renowned racing Camaros were the most-well-known Camaros ordered through the Central Office. The Central Office was in charge of fleet and special orders that required engineering to fill a request. This included fleet trucks and specialty vehicles requiring combinations of parts not included on the list of Regular Production Order (RPO) vehicles. The system used to place orders through this office and the significance it could have on production cars was unknown to the public and also to most car dealers.

In 1968, the Central Office started building special high-performance Camaros. Don Yenko, of Yenko Chevrolet, and Fred Gibb, of Fred Gibb Chevrolet, are known as two of the most influential dealers in getting the COPO Camaro programs in full-swing in 1969. Don Yenko is known to be the man who started the COPO 9561 (cast-iron 427) Camaros, and he had a lot of success selling them, starting at $4,245. Fred Gibb pushed an order through Central Office for 50 1969 COPO 9560 (ZL1 aluminum 427) Camaros and had major problems selling these cars because the aluminum engine raised the base-model price of $2,727 to a staggering $7,200. Sales were extremely slow on the high-stickered 9560 and somehow Gibb was able to get the Central Office to redistribute more than half of the ones he ordered. Only 69 9560s were produced, including the 50 that Gibb ordered. Due to limited production of COPO Camaros, they hold the highest value today.

As with any collector car, its value depends on its condition. If a Camaro is in horrible shape with blowing smoke from loose rings and full of rust holes the size of baseballs in the quarter panels, it's not going to be as valuable as a Camaro in much better shape. On the flip side, a complete COPO Camaro with tons of rusted panels—as if it had laid in a creek bed for 40 years—is going to be worth more than a completely restored RS Camaro, because of its restored-value potential. COPOs are very rare in comparison with the RS Camaro. For true collectors, an unrestored complete car is worth more than a restored car. The reason for this is the buyer has no real clue as to how well the restoration was done or if the parts are really the original parts.

Another highly valued COPO is the COPO 9567, but good luck finding one, since only two of these prototypes were made in 1969 with ZL1 engines and a long list of custom options. Their current existence is unknown.

Double COPO Camaros were 1969s with the COPO 9737 added to the COPO 9560 (aluminum 427) or COPO 9561 (cast-iron 427). The 9737 package was considered to be the "Yenko Sports Car Package" and included a 140-mph speedometer and larger-than-standard sway bar (13/16 inch for 1969, 1 1/16 inches for 1968). The 1969 models also used 15X7 YH rally wheels and E-70-15 WTGT tires. The COPO 9737 was available in 1968, but 1969 was the only year it was ordered along with another COPO, making 1969 the only first-generation Camaro double COPO.

The ZL1 (COPO 9560) is the most well known of the COPO Camaros. There were 69 of these 9560s made in 1969. According to historians, a limited run of 90 aluminum 427 ZL1 engines were built. Only 69 of them ended up in production 1969 Camaros, two ended up in Corvettes, two landed in the COPO 9567 prototype Camaros, and left-over engines (supposedly only 17 more) ended up as factory crate engines. The COPO 9560 wasn't just about the ZL1 aluminum 427 engine, though. It was also equipped with a heavy-duty 12-bolt rear end with 4.10:1 gears, heavy-duty radiator, transistorized ignition, cowl induction, and special front coil springs.

Production by Plant

Production of 1967 Camaros were manufactured as early as August of 1966. Here's the production plant info by year and month.

LOS
1967: August 1966 to July 1967
1968: September 1967 to July 1968
1969: August 1968 to June 1969

NOR
1967: August 1966 to July 1967
1968: August 1967 to July 1968
1969: August 1968 to November 1969

COPO 9561 Camaros had the same options as the 9560, except the engine was the cast-iron 427 (L72), and it didn't have the transistorized ignition. Production numbers were not kept as well as the ZL1-powered COPOs, but the best records show that between 700 to just over 1,000 were produced.

Pace Cars

There's not much documented history on actual production numbers and equipment available on Pace Car Camaros, but there is enough to get collectors spending their money. Of the first three years that the Camaro was built, it was chosen to pace the Indianapolis 500 race in 1967 and 1969. Each time, two specially equipped cars were built to pace the race. There were many other Camaros painted and equipped to look like the actual Pace Cars and they were dispersed to officials and others for promotional purposes. The values of the first-generation Camaro Pace Cars are different, with the highest value placed on cars that actually paced the race. The other Pace Cars and subsequent clones have varying values. The Pace Cars were equipped with small- and big-block Chevy engines.

The 1967 Indianapolis 500 Pace Car is rarer than the well-known 1969 Pace Cars. About 80 1967 Pace Cars were produced. Two cars were built to pace the race, and the others were dispersed to Speedway officials, committee members, and other dignitaries.

The production of the 1969 Indianapolis 500 Camaro Pace Cars was much higher. In 1967, two were specially built for on-track duty at the Indy 500, one was given to the Festival Queen, and about 130 were dispersed to Speedway officials, committee members, and other dignitaries. Later in 1969, Chevrolet produced more than 3,600 replicas for sale to the public.

Vintage Trans-Am Racing Camaros

Camaros with documented vintage Trans-Am racing history are some of the most desirable and valuable first-generation cars. Log books showing true racing history and historic photos really drive up the value of any vintage race car. Without the original log book, these cars have very limited value. Many of these cars that survived were cut up and altered to continue racing through the 1970s, and some were even raced in different classes into the 1980s. Even as altered as these cars have been, they have huge value and can be restored back to their late-1960s racing condition and be worth even more, as long as the restoration is done properly.

Z28 Option

Chevrolet had to retaliate against the Ford Mustang, which was successfully winning in SCCA racing, by producing a Camaro model (Z28) with a 327 block equipped with a

283 crankshaft to equal 302 ci, which met the rule of maximum engine size of 305 ci. The high-winding 302, mixed with suspension, tire, and drivetrain upgrades, made the Camaro extremely competitive, and since Chevrolet produced more than 500 (actually 602 in 1967) Z28s, they were legal for SCCA competition.

Z28 Camaros are loaded with performance options specially picked to make the Z28s more competitive in SCCA Trans-Am competition. Those options included a high-performance 302 engine, dual exhaust, Z28 badging, special front and rear suspension, quick-ratio steering, heavy-duty cooling system, brake upgrades (front disc brakes and in 1969 optional rear disc brakes), upgraded 15-inch wheels and tires, 4-speed transmission, special stripes on hood and rear deck, and more. These cars are very sought after, especially the 1967 Z28, due to low production numbers. It is important to remember that for 1967, no Z28 badging was used; only the classic twin stripes identified the Z28 option. That all changed for 1968. The early cars used only 302 emblems, but soon the classic Z28 emblem replaced the 302 on the fenders. For 1969 the Z28 emblem was standardized.

SS Option

The Super Sport option included a special raised hood with cosmetic inserts and special badges. The SS Camaros received the high-output 350 or a big-block engine and dual exhaust. Many other upgrades made these cars different than the other models, but the upgrades were different from year to year.

RS Option

The Rally Sport (RS) option was strictly an external body-upgrade

Camaros Built

Option	1967	1968	1969
COPO	—	68	769
Z28	602	7,199	20,302
RS	64,842	40,977	37,773
SS	34,411	30,695	34,932
Total	220,906	235,147	243,085

package. It was the most visibly distinctive option package, which featured hidden headlights, different grilles, taillight lenses minus the clear back-up lights (these were relocated to the rear lower valance), special body moldings, and special badges.

Multi-Optioned Camaros

Many times, options were mixed to create personalized and exceptionally desirable cars. For a collector, these multi-optioned Camaros are more valuable than those with only the Z28, SS, or RS options. Camaros could be had with the Super Sport performance package along with the Rally Sport exterior upgrade package to create an SS/RS Camaro (not RS/SS, which is the common misconception). One could also mix the Z28 performance package with the RS exterior package to create the Z28/RS Camaro. The SS and Z28 packages were not mixed because the SS required a 350 or larger engine and the Z28 had the 302 engine.

RPO JL8

Of the options available for Camaros, the JL8 heavy-duty, four-wheel-disc-brake system is probably the most sought-after RPO (Regular Production Order) by true Camaro collectors. Chevrolet developed the JL8 to make the Camaro more competitive in SCCA Trans-Am racing, and it was very successful in reaching that goal. Development on improving the brakes started in 1967 when it was apparent that the current factory disc/drum equipment wasn't going to keep the Camaros competitive.

In 1968, the SCCA Camaros were having success with the "Heavy Duty Service" four-wheel disc packages. For the 1969 model year, Chevrolet offered this awesome brake package as RPO JL8, but killed production of the package in mid-July that year due to lack of sales and production problems. Only 206 JL8-equipped Camaros were produced. The JL8 option added $500.30 and approximately 65 pounds. The price and weight probably kept the drag racers and the average owner away. Heavy Duty Service assemblies were also produced and sold separately but were a little different than the JL8 production units, so original JL8 units are very sought after and easily identified by savvy collectors.

Factory and Dealer Options

There were two types of factory service replacement items: factory-ordered options installed on the production line and factory-ordered parts that the dealer would install.

Instead of ordering a COPO 427 Camaro, a customer could order a small-block Camaro from the dealership and also order a separate engine (like a 427) and have the dealer install the crate engine or do the installation on his or her own. The same kind of scenario went for other specialty Chevrolet "conversion parts," such as the race-bred, dual-carburetor-302 cross-ram intake-manifold system.

Acquiring a Camaro

The Camaro has been an American icon since the first 1967 was

GETTING TO KNOW THE CAMARO

Use letter locations on this 1967 cowl tag with corresponding locations listed on the spreadsheet on the right to decode your tag.

introduced, and over the years it has grown in stature and become a highly desirable commodity. I am an enthusiast of all muscle cars, but, being a complete Camaro fanatic, I probably have a biased opinion when I write the following sentence. Since the year 2000, the most recognizable and sought-after muscle car in the whole world is the first-generation Camaro…at least for Chevrolet enthusiasts. Of the 1967, 1968, and 1969 Camaros, the 1969 is the Holy Grail. It's amazing how its popularity has grown over the years. In fact I'm old enough to have watched the first-generation Camaro grow in popularity from being a good muscle car in the 1980s to a great car to have in your stable in the 1990s. By 2000, the first-generation camaro is more of a status symbol of automotive superiority. It's almost as if having a first-generation Camaro in your garage means you've really succeeded in life…at least as a gearhead.

I can remember a time when entire Camaro shells were cut into small sections and thrown away in dumpsters if a quarter panel had serious rust or if the car was wrecked, assuming it was too far gone to worry about repairing. Or guys would take all the parts off a rusty Camaro shell to build another

1967 Camaro VIN Description

Location	Main data	Code	Description or notes	Location	Main data	Code	Description or notes
A	Build Month	1	January	K	Lower body color	A	Tuxedo Black
		2	February			C	Ermine White
		3	March			D	Nantucket Blue
		4	April			E	Deepwater Blue
		5	May			F	Marina Blue
		6	June			G	Granada Gold
		7	July			H	Mountain Green
		8	August			K	Emerald Turquoise
		9	September			M	Royal Plum
		10	October			N	Madeira Maroon
		11	November			O	Pace Car (NOR only)
		12	December			R	Bolero Red
B	Build Week	A	1st Week			S	Sierra Fawn
		B	2nd Week			T	Capri Cream
		C	3rd Week			Y	Butternut Yellow
		D	4th Week	L	Upper body color: paint or fabric of vinyl or convertible top	1	White fabric (conv)
		E	5th Week			2	Black fabric (conv & coupe)
C	Interior Color	B	Blue			4	Medium Blue (conv)
		D	Red			6	Beige vinyl (coupe)
		E	Black			A	Tuxedo Black
		G	Gold			C	Ermine White
		K	Parchment/Black			D	Nantucket Blue
		R	Bright Blue			E	Deepwater Blue
		T	Turquoise			F	Marina Blue
		Y	Yellow			G	Granada Gold
D	LOS Production Number		Fisher Body code for internal use. Specific details on this code has not been found. Only found on Van Nuys (LOS) built cars.			H	Mountain Green
						K	Emerald Turquoise
						M	Royal Plum
						N	Madeira Maroon
E	Model Year	67	1967			R	Bolero Red
F	Model Type	12437	Coupe STD interior			S	Sierra Fawn
		12467	Convertible STD interior			T	Capri Cream
		12637	Coupe DLX interior			Y	Butternut Yellow
		12667	Convertible DLX interior	M	Group 1 data	D	Power convertible top
G	Assembly Plant	LOS	Los Angeles, CA			E	Tinted windows (all)
		NOR	Norwood, OH			L	Fold-down rear seat
H	Fisher Body production number					X	Power windows
I	Interior Trim	707	Yellow DLX buckets			W	Tinted Windshield
		709	Gold STD buckets	N	Group 2 data	B	3-speed floor shift
		711	Gold DLX buckets			E	Air conditioning
		712	Gold DLX bench			G	Center console
		716	Bright blue DLX bench			H	Heater delete
		717	Blue STD buckets			L	4-speed floor shift
		732	Bright blue DLX buckets			M	TH-400 (LOS only) or PG
		739	Blue STD bench			R	Rear seat speaker
		741	Red STD buckets			S	Rear antenna
		742	Red DLX buckets			U	8-track player
		756	Black STD bench			Z	TH-400 (NOR only)
		760	Black STD buckets	O	Group 3 data	B	Rear defogger
		765	Black DLX buckets			D	Courtesy lighting (coupe)
		767	Black DLX bench			K	Style trim group (Z21)
		779	Turquoise DLX buckets			L	Rally Sport (Z22)
		796	Gold STD bench			S	Interior décor (Z23)
		797	Parchment DLX buckets	P	Group 4 data	F	Remote LH mirror
J	Headrests	H	Bench seat without headrests			K	375HP 396 CID
		T	Bench seat with headrests			L	Z28 302 CID
		Y	Bucket seat with headrests			N	325HP 396 CID
		Z	Bucket seats without headrests			P	SS 350 CID
Conv = Convertible				Q	Group 5 data	B	Rear bumper guards
DLX INT = Deluxe interior						C	Shoulder belts (STD INT)
LH = Left hand						J	3rd rear seat belt
PG = Powerglide transmission						O	Seatbelt delete (export)
STD INT = Standard interior						Y	Deluxe seat belts
						Z	Shoulder belts (DLX INT)

CHAPTER 1

1968 Camaro VIN Description

Location	Main data	Code	Description or notes	Location	Main data	Code	Description or notes
A	Build Month	1	January	I	Lower body color	A	Tuxedo Black
		2	February			C	Ermine White
		3	March			D	Grotto Blue
		4	April			E	Fathom Blue
		5	May			F	Island Teal
		6	June			G	Ash Gold
		7	July			H	Grecian Green
		8	August			J	Rallye Green
		9	September			K	Tripoli Turquoise
		10	October			L	Teal Blue
		11	November			N	Cordovan Maroon
		12	December			O	Corvette Bronze
B	Build Week	A	1st Week			P	Seafrost Green
		B	2nd Week			R	Matador Red
		C	3rd Week			T	Palomino Ivory
		D	4th Week			U	LeMans Blue
		E	5th Week			V	Sequoia Green
C	LOS Production Number	Fisher Body code for internal use. Specific details on this code has not been found. Only found on Van Nuys (LOS) built cars.				Y	Butternut Yellow
						Z	British Green
						Hyphens (- -)	F&SO Paint
				J	Upper body color: paint or fabric of vinyl or convertible top	1	White fabric (CT)
D	Model Year	68	1968			2	Black fabric (CT & HT)
E	Model Type	12437	Coupe			4	Blue (CT)
		12467	Convertible			6	White vinyl (HT)
F	Assembly Plant	LOS	Los Angeles, CA			A	Tuxedo Black
		NOR	Norwood, OH			C	Ermine White
G	Fisher Body production number					D	Grotto Blue
H	Interior Trim	711	Ivory&Black DLX buckets			E	Fathom Blue
		712	Black STD buckets			F	Island Teal
		713	Black STD bench			G	Ash Gold
		714	Black DLX buckets			H	Grecian Green
		715	Black DLX bench			J	Rallye Green
		716	Ivory Houndstooth DLX buckets			K	Tripoli Turquoise
		717	Blue STD bucket			L	Teal Blue
		718	Blue STD bench			N	Cordovan Maroon
		719	Blue DLX buckets			O	Corvette Bronze
		720	Blue DLX bench			P	Seafrost Green
		721	Gold DLX buckets			R	Matador Red
		722	Gold STD buckets			T	Palomino Ivory
		723	Gold STD bench			U	LeMans Blue
		724	Red STD buckets			V	Sequoia Green
		725	Red DLX buckets			Y	Butternut Yellow
		726	Turquoise DLX buckets			Z	British Green
		727	Turquoise DLX bench			Hyphens (- -)	F&SO Paint
		730	Parchment&Black DLX buckets				
		749	Black Houndstooth DLX buckets				
CT = Convertible top							
DLX = Deluxe interior							
F&SO = Fleet and Special Order							
HT = Hard top							
STD = Standard interior							

Use letter locations on this 1968 cowl tag with corresponding locations on the spreadsheet on the left to decode your tag.

Camaro and store the parts they didn't need, then take the shell to be recycled as scrap metal. Nowadays, these once "lost causes" are being pieced together as project cars worth saving, but some of the most seasoned sheetmetal veterans question the authenticity of these body parts.

The collectability of first-generation Camaros for enthusiasts and "true" car collectors has turned the Camaro restoration business from a couple-million-dollar industry into a multi-billion-dollar industry in a few decades. That doesn't include the huge performance aftermarket industry helping drive the booming first-generation Camaro parts industry even further through the roof, as well as the values of the cars. By 2005, DII had started to release its completely new first-generation Camaro shells. This drove the price of basket-case Camaros, especially 1969s, with only a recognizable cowl panel with a trim tag and VIN tag, to a new high. Basically, if a guy scrapped a whole basket-case 1969 shell, he could sell the trim and cowl tag for a minimum of around $2,000 in 2008, since the tags can now be added (illegally) to the new stamped shell to make the new body a registered 1969 Camaro. The prices for original first-generation Camaros have risen to new highs since then,

GETTING TO KNOW THE CAMARO

Use letter locations on this 1969 cowl tag with corresponding locations listed on the spreadsheet on the right to decode your tag.

making them out of reach for many entry-level enthusiasts.

To confirm it's an original RS, look for signs of tampering, such as behind the back-up lights. Most people get in a hurry and don't spend the time to perform a good job when cutting the holes for the backup lights in the tail panel. A good job looks factory, with good straight lines for the lens and a clean round hole through the inner panel for the electrical wire to pass through. There are two panels to cut through in order to mount the lights: the outer tail panel and the inner tail panel. Remove the back-up lights and inspect both holes for signs of being non-original. Of course, the signs of not being original will be there if the whole tail panel has been replaced, so inspect the inside of the trunk for signs of tail-panel replacement before spending too much time inspecting the back-up lights.

A tell-tale sign of tail-panel replacement is the lack of seams where the tail panel and quarter panels meet. Body shops may unknowingly fill the seams with body filler, which is a dead giveaway of a poorly done repair. Another sign of repair is the absence of the distinctive seam where the quarter panels meet the fill-panel that spans the trunk between the back glass and the deck

1969 Camaro VIN Description

Location	Main data	Code	Description or notes	Location	Main data	Code	Description or notes
A	Model Year	69	1969	G	Upper Body Color	A	White convertible top
B	Model Type	12437	Coupe			B	Black vinyl or convertible top
		12467	Convertible			C	Dark blue vinyl top
C	Assembly Plant	VN	Van Nuys, CA			E	Parchment vinyl top
		NOR	Norwood, OH			F	Dark brown vinyl top
D	Fisher Body production number					S	Midnight green vinyl top
E	Interior Trim	711	Black STD buckets			10	Tuxedo Black
		712	Black DLX Buckets			40	Butternut Yellow
		713	Black Houndstooth DLX buckets			50	Dover White
		714	Yellow Houndstooth DLX buckets			51	Dusk Blue
		715	Dark blue STD buckets			52	Garnet Red
		716	Dark blue DLX buckets			53	Glacier Blue
		718	Medium blue STD buckets			55	Azure Turquoise
		719	Medium blue DLX buckets			57	Fathom Green
		720	Orange Houndstooth DLX buckets			59	Frost Green
		721	Medium green STD buckets			61	Burnished Brown
		722	Medium green DLX buckets			63	Champagne
		723	Midnight green STD buckets			65	Olympic Gold
		725	Midnight green DLX buckets			67	Burgundy
		727	Ivory&Black STD buckets			69	Cortez Silver
		729	Ivory Houndstooth DLX buckets			71	LeMans Blue
F	Lower Body Color	10	Tuxedo Black			72	Hugger Orange
		40	Butternut Yellow			76	Daytona Yellow
		50	Dover White			79	Rallye Green
		51	Dusk Blue			Hyphens (- -)	F&SO Paint
		52	Garnet Red	H	Build Month	1	January
		53	Glacier Blue			2	February
		55	Azure Turquoise			3	March
		57	Fathom Green			4	April
		59	Frost Green			5	May
		61	Burnished Brown			6	June
		63	Champagne			7	July
		65	Olympic Gold			8	August
		67	Burgundy			9	September
		69	Cortez Silver			10	October
		71	LeMans Blue			11	November
		72	Hugger Orange			12	December
		76	Daytona Yellow	I	Build Week	A	1st Week
		79	Rallye Green			B	2nd Week
		Hyphens (- -)	F&SO Paint			C	3rd Week
						D	4th Week
						E	5th Week
DLX = Deluxe interior				J (NOR only) started approx. Nov. '68 production	Exterior Trim	D80	Front and Rear spoilers
F&SO = Fleet and Special Order						X11	Z21 style trim
STD = Standard interior						X22	Z21 style trim SS396 equipped
						X33	Z21 style trim Z28 equipped
						X44	Non Z21, Z28, & SS
						X55	SS350 equipped non Z21
						X66	SS396 equipped non Z21
						X77	Z28 equipped non Z21
						Z10	Coupe w/Indy Pace Car features
						Z11	Convertible Indy Pace Car
				J (VN only)	VN Production Number (previously LOS)		Fisher Body code for internal use. Specific details on this code has not been found. Only found on Van Nuys (VN) built cars.

CHAPTER 1

If you're looking at a Camaro without a VIN plate, you should probably keep your money in your pocket and move to the next car. This 1968 Camaro was also missing its cowl tag. With the hidden VIN numbers you could reassemble the original VIN and check the numbers with your local law enforcement to ensure the car isn't stolen.

lid. If a quarter panel has been replaced, there may also be visible trauma inside the trunk and poor transition work in the rain gutter under the trunk lid.

Poorly executed panel replacement lowers the value of the car, because the cost of fixing it can be very expensive. We've seen replacement quarter panels welded in place 1/4 inch too far forward. The door would never align properly, and the problem was not noticed until an accident caused a 1/4 inch of body filler to fall off the end of the quarter panel. If you're looking at purchasing a car with new weld-in body panels, such as a tail panel, quarter panel, rocker, floorpan, etc., consider the work that's been done. If body filler and primer have been applied, you won't know if the bodywork was properly completed until paint is applied and you drive the car around for a while, but by then it's too late. Sometimes you can see evidence of bad bodywork by looking at a panel from the back side, but this isn't always possible or the work quality is not immediately apparent.

If you come across a Camaro missing its VIN and cowl tag, you should probably take your money and run. If the car is a really sweet deal, you need to get the hidden VIN from under the cowl panel and run it through the department of motor vehicles or system at your local police department to confirm the car is not stolen. The last eight digits of the hidden VIN are the most important. With minimal detective skills you can determine the first five numbers. The first two digits are going to be "12." The third digit is the toughest one, but you can look for signs of the car possibly having come with a 6-cylinder or a V-8 by fuel line size, engine info sticker on the radiator core support, or in a rare case you find the broadcast sheet. If you can't figure out the third digit, have the police run the VIN both with a 3 for 6-cylinder and a 4 for V-8, to see if either are stolen. The fourth digit is easy; it's either a 3 for hardtop or a 6 for convertible. The fifth digit is always a 7 on these Camaros. If the hidden VIN has been removed, you need the broadcast sheet to determine the VIN.

Check on Common Rust Areas

The first-generation Camaros weren't dipped in a chemical rust-proofing tank before they were painted at the factory like newer cars. They can rust very badly if neglected or driven in the snow on a regular basis. These cars are more than 40 years old, and almost all of them have some visible rust somewhere (or they've been previously repaired). It is possible, but rare, to find a car that has been raced all its life and stored in a garage during its off time.

Camaros typically rust in the area around the bottom of the rear window because water pools up in the poorly designed window channel without a way to drain out. Mix in a little trapped dirt and moisture, and rust is going to spread in a hurry. Rust holes are also visible on the metal interior panel. The trunk pan was full of rust because water flows right into the trunk.

The common rust areas are in the gutter around the rear window under the stainless trim (in most cases it's visible with the trim installed), the trunk pan (because the rear window rust lets water collect in the trunk), and the quarter panel behind and in front of the rear tire. In addition, rust often forms around the rain gutter around the side glass, the bottom of the front fender behind the front tire, the top of the dashboard inside the car at the base of the windshield, the roof panel above the windshield, and just about any floorpan. There are other areas in the cowl area, hidden by the front fenders that you can't see until the fenders are removed.

Visible rust holes on a car are usually the tip of the iceberg because you can only see about 10 percent of the actual rust, so you should be very careful when purchasing a car with substantial rust. If you see a 1/2-inch hole in the quarter panel, the inner fenderwell and other adjoining panels are also possibly damaged by rust.

If you see a car with thin paint on the top horizontal surfaces

Rear quarter panels rust like this in front of and behind the rear tire. The overlapping metal panels in the wheelwell trap dirt. Just add water flying off the tire between those panels, and rust begins to form.

Rain gutter panels are supposed to be sealed to the body with seam sealer. Over the years, the seam sealer can crack and allow moisture to get between the A-pillar post and the rain gutter panel and breed rust.

(trunk, hood, roof, etc.) and maybe even a tiny bit of surface rust, there's a good chance that the car still has its original single-stage lacquer paint. If none of the other panels show signs of different shades or finishes, it's a good sign that there might not be repairs hiding a troubled past. Signs of repainted and patched paint can hide a myriad of serious rust and panel replacement that might not have been performed professionally. (More on rust in Chapter 3.)

Finding Aftermarket Parts

Before even looking for a Camaro, have a plan for what you're going to do with it. Performing a full concours restoration and building a hot rod are two completely different approaches, and therefore you will consider purchasing different cars. If you look at a car that's missing half of its small part (bumpers, brackets, wiper assembly, gauge cluster, headlight assemblies, factory bolts, etc.) and you want to put it back to stock, you're going to spend many hours at swap meets and salvage yards scrounging for parts. If you're building a hot rod or a racing car, you may decide to install aftermarket gauges, fiberglass bumpers, reproduction brackets, and hardware-store bolts to replace missing items.

The Camaro restoration parts manufacturers are making almost every part you could ever need, but there will always be some part you need they don't offer. With reproduction parts, there are different levels of quality on the market, and this is especially true when it comes to replacement quarter panels. One company may sell a quarter panel at a very affordable price, but the stamping doesn't have crisp lines, and it requires an additional 30 hours to get it to fit well. On the other hand, a competing company may sell a more expensive panel, but it has crisp lines and fits much more precisely. Therefore, it requires minimal work to get it to fit the tail panel and trunk jamb. Although a part may be less expensive, it often requires much more work to properly fit, and the time invested to install the cheap quarter makes it much more expensive once all the work has been completed.

The best scenario for purchasing a project Camaro is to buy one that has most or all the parts you are going to use—in good shape. If a project car has all the parts you want, but each part is broken or damaged, you may want to look for a different car to start with. It's better to pay a little more for a good start than pay less to start, which can turn into spending excessive money in the long run. These projects are expensive enough. Make sure you invest in a good foundation. Scrounging the swap meets for one ridiculous bracket or small part gets expensive and tiring.

Keep Investment Value in Mind

If you're buying a car to restore for concours restoration, you must consider buying one of the most desirable models, so you have a better return on your investment. If you turn a factory-equipped 6-cylinder Camaro into an RS, SS, or Z28 clone, it will never have the same value as if you performed the same work on a factory-original RS, SS, or a Z28 Camaro V-8, because the third digit in the VIN will still show the car came equipped with the L6. Therefore, spend your money wisely if you're going into your restoration as an investment. Also, keep in mind that there are 1967 and 1968 Camaros, as well as Firebirds, with 1969 Camaro panels on them. Check the VIN to be sure you're not buying a 1968 Firebird somebody converted into a Camaro. These incorrect cars are out there on the market and have very diminished resale value.

To buy a good candidate for restoration, you want to purchase a good unrestored and unmolested car with as little rust or physical damage as possible. Partially restored cars without finely detailed pictures documenting the restoration process are worthless because it's really easy to build a clone and claim it's a real SS, RS, or Z28. If there's no broadcast

CHAPTER 1

sheet or window sticker to accompany a car, you're simply relying on the word of the owner, along with the parts on the car, to determine if it's truly what the owner says it is. A car has more value if it's obvious the car hasn't been tampered with and has all the specific equipment to prove it's truly an original SS, RS, Z28, or big-block car. This is a two-way street, so before you start a restoration, keep photographic and written documentation of each step of the process. This increases the value of your car if you ever sell it.

Identifying a Fake

Nobody likes to get duped, and many unsuspecting buyers have purchased an SS, RS, or Z28 clone when they thought they were buying an original example. Many automotive enthusiasts are looking for a car with good investment value. There's not much worse than paying extra money for a car that

True Tourtillott's Camaro was painted pink from the factory. The owner's grandfather can attest to it because he bought the car brand new from Courtesy Chevrolet in San Jose, California, in 1969. Special-order paint alone did not mean the car was a COPO, it simply meant that the Fleet & Special Order Department was involved.

isn't what a seller says it is. For instance, you may pay an additional few thousand dollars for a 1969 SS than you'd pay for a Plain Jane 307-ci-powered 1969 Camaro of the same exact condition.

The most obvious fake is a V-8 car with the VIN plate designating it was equipped with an L6 6-cylinder engine, when the car is being sold as a V-8 car.

The 1969 is a fairly easy car to identify as a possible SS or Z28 fake because all SSs and Z28s had performance engines that, in turn, required a dual-exhaust system, which, in turn, had a special reinforcement plate welded to the side of the rear frame rail behind the driver's-side rear tire. Of course, somebody could simply weld on a reproduction plate, but some unscrupulous sellers may overlook it. The 1967 and 1968 have something very similar, but easier to fake. There are two holes in the same spot on the side of the rear frame rail that have factory nut-serts installed if it was a factory dual-exhaust car. Nut-serts are fairly easy to install on the 1967 and 1968, but the 1969 plate is a little harder to fake.

Decode Your Camaro

You can successfully decode some basic information on what options your car had when it rolled off the assembly line. Simply looking at the equipment bolted to your car means almost nothing because building "clones" has always been a big business, and some dishonest car sellers know how to trick even the most knowledgeable Camaro enthusiast. To find more detailed information about specific options, you need knowledge and skill to identify small details. The easiest way to decode your Camaro is with the original window sticker,

broadcast sheet, and Protect-O-Plate, but hardly anyone is lucky enough to find these with the car.

Broadcast Sheet

How do you get documentation for your first-generation Camaro to prove its originality? Well, the only way to know for sure is to find the broadcast sheet (also known as the build sheet)—and that's often a long shot. Usually, someone has removed it from your car sometime between the day your car was originally purchased back in the late 1960s and today. I've heard of them being located in multiple places in the car. The most common location is under the rear seat, sandwiched between the fuel tank and the trunk pan on the driver's side—barely seen by looking into the fenderwell…if it's still there and didn't completely deteriorate. I've also heard of them being located under the dash. There are also some cars with low miles owned by original owners that never came with a broadcast sheet, so not all cars come with the hidden paperwork.

Camaro historians say that they never expect Chevrolet to find lost records of option documentation for the 1967, 1968, or 1969 model years. Instead, we just have to use the current method of breaking down the VIN, decoding the cowl tag, hoping we find a broadcast sheet (which hardly ever happens anymore), and crossing our fingers that we can find an untouched Camaro entombed in a moisture-proof vault in the back of some old farmer's barn in the California desert. We also have to abide by the current honor system that if somebody says they have an original SS/RS Camaro, that it really is one (unless proven otherwise by the VIN showing it was a 6-cylinder). I have a feeling that a lot of Camaro owners out there would be unhappy if Chevrolet ever did mysteriously find the lost production records.

COPO F&SO

Fleet and Special Orders (F&SO) were designated on the broadcast sheet in the same box as COPO, but the two were separate entities. The COPO was reserved for special fleet modifications, which were mostly reserved for truck and larger cars that received government, law enforcement, and taxi packages requiring special engineering in order to make sure all the parts required would work together. Restoration experts claim the F&SO Department offered parts and accessories that were not standard options, but didn't require engineering to carry out the order, which included special-order paints. Basically, if your Camaro was ordered with special-order paint (designated with double dashes or blanks in the "paint" section on the cowl tag), your car is not considered a COPO car, but it was a special-order vehicle. Some dealers wanted a distinct-colored Camaro to make the it stand out from Camaros on other lots, so they would special order colors not available in the standard color offerings.

Identifying Your VIN Plate

The 1967 Camaro VIN plate is located in the doorjamb and is held in place by distinct rosette rivets.

Starting in 1968, all manufactured automobiles had to comply with new federal government regulations requiring all automobiles to have a vehicle identification number visible in plain view from outside the vehicle. That meant, starting in 1968, the Camaro had its VIN plate moved to the left side of the dashboard, so if you were a police officer standing in front of the driver's door, it could be seen through the windshield.

The same rosette rivets used on the 1967 VIN plate were used on the 1968 and 1969 Camaro VIN plates, but they were hidden under the dash pad. In 1969, Chevrolet added a conformance VIN sticker on the driver's-side door in the doorjamb above the striker. This required an entirely different door frame, which had two raised lines stamped into the door to locate the sticker. Chevrolet also hid two VIN numbers stamped on the body, which only contain the last eight digits. To confirm that your Camaro is exactly what the VIN plate says it is, you can compare the VIN plate numbers to the two hidden VINs. If the car has been stolen often the VIN plate has been swapped. In other cases, to clone the car into a more collectible model the VIN number has also been changed.

The first hidden stamping is located on the top of the cowl under the cowl vent panel; in some cases you can see the number through the grille covering the cowl openings. The other hidden VIN is stamped on the front of the firewall and located behind the heater duct or the air-conditioning (A/C) duct. Removal of the duct is necessary in order to see the stamped numbers. John Hinckley from the CRG says these were stamped by an air-operated stamping device with a back-up anvil that folded behind the panel to reduce distortion in the panel. These numbers were stamped after the body was welded together and painted. Because an assembly-line worker was in charge of using this stamping device, some mistakes happened in the process. In some cases the number was not stamped and in other cases the number was stamped incorrectly.

In an extremely rare case, the VIN in one or both of the hidden locations was stamped incorrectly at the factory. You can see the factory worker stamped this one incorrectly on the assembly line. A factory worker later restamped the 7 as the correct 9. Since it's two digits off, one can only wonder how many cars got the wrong stamps that day.

The last eight characters on your VIN plate (in the driver's doorjamb on 1967s and on top of the driver's-side dash on 1968s and 1969s) should match the numbers hidden in these two locations. If they don't match, somebody switched the VIN plate illegally and it's possible the car is a stolen vehicle. This stamp was done by a person, so it's always possible that it was missed.

CHAPTER 1

The VIN specifies the particular car's engine, year, and the plant where it was manufactured. A significant number on the tag is in the third position, which signifies 6-cylinder or V-8 engine. Some people may think the VIN signifies the order in which the car was produced, but that order was designated by the BODY number on the cowl tag.

The stamp (shown upside down) on the top of the cowl can sometimes be seen through the cowl panel without removing it. The stamp on the firewall can't be seen unless you remove the engine compartment heater-box ducting.

If the two hidden VIN stamps aren't present, there may be several common reasons: The VIN was never stamped, somebody has replaced these panels because they were extremely rusted out, or the body you have is a completely new stamping from Dynacorn and somebody attached an old VIN plate.

What the Numbers Mean

The VIN gives some information about the particular car's engine, year, and the plant where it was manufactured. A significant number on the tag is in the third position, which signifies 6-cylinder or V-8 engine. Some people may think the VIN signifies the order in which the car was produced, but that order was designated by the BODY number on the cowl tag.

- The first digit is 1, which denotes Chevrolet.
- The second digit on the Camaro is 2. It denotes F-body, which could be Camaro or Firebird, but since the first digit is 1, it would denote Camaro.
- The third digit shows the engine type; 3 = L6 or 4 = V-8.
- The fourth digit denotes body type; being 3 = hardtop or 6 = convertible.
- The fifth digit is a 7, which simply equates to a two-door couple.
- The sixth digit denotes the model year; 7 = 1967, 8 = 1968, or 9 = 1969.
- The seventh digit is a letter designating the assembly plant; "N" = Norwood, Ohio, or "L" = Los Angeles/Van Nuys, California.
- The eighth through the thirteenth digit are the actual sequential production unit number.

Deciphering the Cowl Tag

The cowl tag is also known as the Body Number Plate (according to Fisher Body service manuals), Trim Tag, or Trim Plate. Because there's a slim chance of finding the paper broadcast sheet, the cowl tag is a much more reliable source of information because the stamped-aluminum plate is much more durable than steel. The 1967 cowl tag shows a lot more about the options that a car was originally designated to have when it left the factory; but, it could be wrong in some cases, because a special-order option could have changed after the tag was stamped.

The 1968 and 1969 tags were more vague and don't help restorers identify much about their vehicles.

Instead of taking the option codes stamped on the tag for gospel, you should know that the tag has the information that the Camaro was slated to be built with. Unfortunately for restorers, special-order options and paint could have changed, but the cowl tag would not have been updated.

Body Style

The cowl tag had 7 ST (body-style code) digits:
- The first two digits were the model year; 67 = 1967, 68 = 1968, and 69 = 1969.
- The third and fourth digits were always 12.
- In 1967 the fifth digit designated interior style; 4 = standard interior and 6 = custom interior.
- In 1968 and 1969, the fifth digit was always 4.
- The sixth and seventh digits designated body type; 37 = coupe and 67 = convertible.

Paint codes on cowl tags that are blank or with two dashes (- -) usually denote a special-order paint color.

A dash followed by a number in 1967–1968 or a letter in 1969 indicates a vinyl top. For example, –B denotes special paint with a black vinyl top in 1969.

Production Week

The production week on the cowl tag is the week the actual cowl tag was produced, but it was not the day or week that the body was assembled. Research shows that 1969 Camaros built in June were stamped 6A. Even if Norwood (the larger producer) alone ran 24 hours per day, it could not produce all 13,000 Camaros in the first week of June.

Body Number

To the left of "BODY" on the cowl tag is the "body build number" or "unit number," which was assigned by Fisher Body (the division in charge of welding the actual body shell together). The body numbers were assigned in sequence as the bodies were built. In Norwood, the sequence included all the bodies by each model year. The first NOR body produced for the model year would be assigned number 1. In Los Angeles, the body number was in sequence by the body and interior type. The first LOS standard interior coupe was assigned number 1, and the first LOS standard interior (convertible) was assigned number 1. This format was also used for deluxe interior coupes and convertibles, so there are four different LOS-built Camaros and each year has a body number of 1.

Interior Trim

To the right of "TR" on your cowl tag is the interior trim designation. Depending on the year of your first-generation Camaro, the numbers and letter denoted interior color, standard or custom panels, front seat (buckets or bench), and convertible top color (1969 only).

Paint

The cowl tag has "PAINT" stamped on the right side. It designates the color information of the car when it rolled out of the factory (it specifies paint as well as vinyl interior and convertible top color). If the car was slated to be a special-order paint or a "stripe delete" color, the cowl tag would have a couple of dashes or blanks in place of the color code digits. On rare two-tone Camaros built in 1969, two paint codes were used. For example, 53-50 specifies Glacier Blue with Dover White painted top. These rare cars used the standard vinyl-top moldings as the break point for the paint. The first numbers always designates "lower paint" and the second number designates upper or roof paint.

Camaro O'Canada

GM of Canada was good at keeping Camaro option records. Canadian-sold Camaros were well documented, so if your Camaro was originally sold in Canada, you may be in luck. GM vehicles sold after 1964 have been well documented. You can find the original location of the new vehicle sale, production date, interior and exterior color, type of engine and transmission, and the vehicle options. To get detailed vehicle information check out www.vintagevehicleservices.com.

Plants

The Norwood, Ohio, plant produced about 70 percent and Los Angeles, California, plant produced about 30 percent of all the 1967 to 1969 Camaros. The Camaros produced at the Norwood plant are easily identified on the VIN plate with an "N" in the seventh position from the left, and the cowl tag was stamped "NOR." The Los Angeles (Van Nuys) Camaros are identified by the "L" stamped in the seventh digit from the left. The cowl tag was stamped "LOS" until late 1968 when it was replaced by "VN."

Most Camaro enthusiasts are very aware of these two U.S. plants, but are surprised to learn that other plants also produced the cars. The Camaro was produced outside of the United States in factories and assembly plants controlled by GMOO (General Motors Overseas Operations) and GMODC (General Motors Overseas Distribution Corporation). These facilities may have received Camaros in subassemblies with most assemblies already complete in SKD (semi-knocked-down) kits, which were believed to have been assembled in a Venezuelan factory to save money on taxes and port fees. These facilities may have received Camaros in CKD (complete-knocked-down) kits. The cars were sent completely disassembled, except for the engines, rear axles, and other small assemblies. In addition, the bodies were shipped in pieces and welded together in their designated country's factory, such as the Philipines, Switzerland, and Belgium. Research shows that all these SKD and CKD kits came from the Norwood, Ohio, plant.

In 1967, the VIN plate was attached to the body in the driver's doorjamb, closest to the hinges. These were attached with distinctive Rosette rivets.

The laws changed for the 1968 model year, which required all VIN numbers to be visible from outside the vehicle without obstruction, so the plate was moved to the top of the driver's-side dashboard where it was visible through the windshield.

CHAPTER 2

GETTING STARTED

The first thing you need to do is figure out if you're going to tackle all the work yourself or if you're going to hire a qualified shop to do it for you. Not everyone who has the ambition to restore a car has the skills to do it. So, you have to assess your skills, available time, and access to tools. But, more importantly, there are very few people who are highly skilled in every area of restoration—engine rebuilding, suspension, drivetrain, bodywork, painting, interior, etc. Inevitably, you'll have to farm out some aspect of a complete restoration. Before you send work out, consider that you won't learn new skills without going outside of your skill set or comfort zone.

Many daunting tasks pop up during the restoration process. It's great if you can say that you built your entire car, but you have to perform some soul searching and be wise about performing certain tasks. If you have reservations about rebuilding your engine, transmission, or rear axle, you may consider sending these out to shops equipped with the necessary tools and skills. It takes a wise person to put ego aside and know your limitations and pay somebody else to perform work that you feel is a little over your head. There comes a point where clearances are too critical to take chances with. Missing a small detail can be very costly. Bodywork and paint also require skills that not everyone has the ability or knowledge to perform. Some tasks require a great deal of specialty tools that you may use once in your lifetime or are too expensive to fit in the budget. Think of tools you do purchase as an investment for future projects down the road.

Start each restoration by documenting the car and parts you are restoring. Since digital cameras are so affordable these days, taking plenty of photos of your car is a very good practice. The pictures are often indispensable reference sources when you reassemble the project. In addition, you can share detailed images with people who may be able to help you through a tough spot. Image records also add value if you ever sell the car.

Skills

For good final results, attention to detail and the aptitude to do well are critical. Unless you're a seasoned

GETTING STARTED

professional, there is a learning curve for getting a great final product. Some tips on how to perform some tasks are covered in this book, but others are learned skills that can't be described or instructed through words and photography.

Experience is the best skill to acquire, which means if you come upon a task of welding, painting, removing dents from trim, etc., you should practice on a test piece first. Performing a new task on a salvageable part without experience is asking for trouble. If you can't get the knack of the task, this may be time for you to suck it up, put your pride in your pocket, and have a trained professional do the job. It takes a knowledgeable person to perform a task, but it takes an intelligent person to know his or her limitations. Don't let pride get in your way. Pride can get expensive.

Tools and Tool Basics

What tools are required to perform a great restoration? Basic hand tools such as wrenches, sockets, ratchets, screwdrivers, etc., are required to perform most of the tasks of a restoration. A big help in restoring a car is knowing what tools are the correct tools for the job. Simple knowledge about the differences between No. 1, 2, 3, and 4 Phillips screwdrivers, for example, can save you hours of grief due to stripped-out screws and damaged screwdrivers. There are also multiple sizes of slotted (a.k.a. flat-blade) screwdrivers. If a screwdriver doesn't fit snuggly in the head of the screw, get the right tool before applying torque. If you only own metric tools, you can get by in some cases, such as using a 13-mm socket to remove a 1/2-inch nut, but it's not worth the times when a bolt or nut strips out because it was a little bit of a loose fit. Get a set of standard tools before starting your restoration.

The benefit of simply restoring a Camaro with stock parts is that all the parts bolt on (at least they are supposed to), especially if they are all original stock parts. When you start adding non-stock parts, you need additional tools to make custom parts or at least modifications to parts to make them fit or attach to your Camaro. You also need some additional tools (which can be rented) to perform a restoration such as a coil spring compressor, engine hoist, etc. A good rolling floor jack is required. If you don't have a hydraulic press to press ball joints and bushings into control arms, obviously, you have to take these jobs to a local shop.

A capable air compressor is a big help for most car projects and a necessity for a complete restoration project. The compressor is worth its weight in gold to get debris out of areas you can't otherwise reach and to remove dust from sanded areas. If you're shopping for an air compressor, make sure it fits your needs as far as CFM and tank size. You can get a dual-action (DA) sander for sanding the body. Other nice air tools to have are an air ratchet, cut-off wheel, drill (with a wire wheel for cleaning), and high-speed grinder. If you plan to use a DA sander or cut-off wheel, you're going to need a compressor with at least a 50-gallon tank and about 10 cfm at 90 psi. You can get away with less, but you're going to have very short bursts of work and long waits for the pressure to build, along with many hours of cursing.

Materials and Specialty Tools

For restoration purposes, you're going to need a lot of materials and some specialty tools. The cost of these items should be figured into your budget. If you're in a car club or have friends performing some of these tasks, you might be able to split the cost with them. Each job requires its own list of items. For instance, when replacing a quarter panel, you need a spot weld cutter, a welder, grinding discs, paints, fillers, cut-off wheels, etc. Installing front sheetmetal simply requires body shims and simple hand tools. The jobs requiring the largest materials are the ones dealing with bodywork and paint, which is why it sometimes makes the most sense to have a shop do most of this type of work.

Building an engine is another job that requires a few expensive specialty tools. Clearances between bearing and machined surfaces are critical and require a few tools you'll use again only if you build more engines in the future.

Other specialty materials needed are polishing supplies if you're refinishing trim pieces; adhesives for weatherstrips, windows, and door assemblies; and special lubricants for window tracks; etc.

Where Do I Start?

I've had a few people shake my hand and tell me that the following was the best advice they had ever read, because it paid off for them in a big way! Here it is: You don't have to tear your car apart down to the last nut and bolt in order to restore it, unless you're going to perform a full rotisserie, frame-off, concours restoration. If you're simply restoring your car to enjoy driving it on a regular basis, in most cases, you do not have to completely disassemble the car. You can perform most of the restoration procedures covered in this book as smaller separate projects. But you need to decide how far to tear your car down before you remove a single bolt. Keep in mind that 70 percent of frame-off projects are sold before they are finished, and over half of those are sold as completely disassembled cars in dozens of boxes. Typically, a car sold as a shell and a bunch of boxes is sold for less than the original purchase price. An assembled car in a state of disrepair is worth more than a car in boxes. You need to keep that in mind if you're tackling a frame-off project as more of an investment than a fun car.

Overwhelmed by the magnitude of the work, many at-home restorers lose interest in a restoration project. When all the parts come off of the car in a hurry, it can give you the false impression that they will go together at about the same speed, which couldn't be further from reality. The new or reconditioned parts they have to be adjusted, resealed, aligned, lubricated, trimmed, fitted, etc.

Repair or Replace?

When performing a restoration job with the goal of doing it correctly and well, assume the job will take three to four times longer than you anticipate. Nothing simply bolts into place and fits perfectly, especially if you are using multiple reproduction parts. As hard as reproduction manufactures try (some try much harder than others) to make parts identical to the factory part, it's tough to beat the fit of a factory-original part. Not even NOS parts fit as well as originals, especially when it comes to sheetmetal. You need to recognize that when purchasing new parts. Are you replacing the part because it's missing, or is it simply not in great shape? Does the part you're replacing really need to be replaced? Can you repaint it or repair it? I've personally repaired two broken corners on the plastic gauge cluster on my 1968 Camaro with Super Glue and body filler. Luckily, I had all the pieces. It fits perfectly because it was an original factory-produced part and I saved the $125 for a new gauge cluster, plus shipping. It's been perfect for 13 years and 25,000 miles.

You should also save all original stainless and aluminum trim. There's no substitute for original molding and trim. Reproduction trim just doesn't fit as well as the original trim. Even if your original trim is damaged (within reason), a trained repair specialist can save it. A lot of defects, scratches, and dents can be pounded and polished out.

Game Plan

Don't go into a restoration without a plan. Basically, if you fail to plan, you plan to fail. Sure, you can successfully restore a car without a plan, but you do it at the expense of more time and money. You're probably reading this section for some advice on a system that benefits your project. Search deep inside your heart and figure out what you really want out of your Camaro when it's all done.

If you want a concours restoration to be judged against other restored Camaros at Chicago Muscle Car and Corvette Nationals or Pebble Beach, assume you'll have to acquire almost unobtainable parts and perform tons of research to know all about the small tedious parts the judges look for—the correct production tag for your radiator, correctly dated glass, etc.—and you can assume a full rotisserie restoration is going to be necessary.

So, if that's your goal, you need to start with a good unrestored candidate with almost all of its original parts, and every part needs to be removed for repair or refinishing. If your goal is to restore a clean weekend driver, a nice hot rod, or a Pro-Touring street machine, you really need to have an end goal before removing one nut or bolt.

Sure, you can change the plan along the way, but if you make a plan and stick to it, you reduce the additional time and money spent performing tasks more than once and also reduce final project costs. If you're even the slightest bit on the fence, talk to multiple Camaro owners who own cars in the same format you may want to build your car into. Pick their brains about what they like or don't like about their Camaro. This information may help you figure out what you really want to do to your Camaro.

Make a list of the parts needed to perform your restoration project, whether it's a full restoration or a smaller project. Creating a list for a full restoration is a difficult task, not just because the list would contain more parts, but because it may contain projects you need to farm out to shops. Even if you're a fairly skilled mechanic, there will probably be a few tasks to send out. Often, the typical enthusiast restorer does not have the specialty equipment or the specific skills required. Therefore, he or she will probably outsource the work to a professional shop that takes an extended period of time to complete. And it's not that these tools cannot be acquired or certain skills learned,

GETTING STARTED

Steps in the Process

Here are the installation procedures for doing the paint and bodywork first (including interior and trunk paint).

1. Completely assemble doors.
2. Install doors.
3. Paint frame, and bolt to body.
4. Install front suspension and steering linkage.
5. Install steering column.
6. Install rear suspension.
7. Install brake master cylinder.
8. Install brake and fuel lines.
9. Install body wiring harness.
10. Install all dash and climate-control accessories.
11. Install heater box and blower motor on firewall
12. Install fuel tank.
13. Install engine and transmission.
14. Install radiator and radiator-core support.
15. Install wiper system.
16. Install interior (headliner, rear package tray, carpet, seat belts, quarter side panels, rear seat, front kick panels, front seats, door panels).
17. Install front and rear glass.
18. Install lower stainless front windshield trim.
19. Install and align doors to quarter panels and rocker panels.
20. Install emblems on fenders.
21. Install front inner fenderwells.
22. Install and align fenders, header panel, lower valance, cowl vent cover, and hood (in that order).
23. Install front bumper.
24. Install rear bumper and taillight assemblies.
25. Install headlight assemblies and grille.
26. Install engine and front lighting electrical harness.
27. Install remaining body and windshield trim.

but it's often not realistic to acquire them when you're restoring an entire car. These specialty skills are: engine machine work, paint, bodywork, and panel replacement.

From personal experience, machine shops and body shops typically take two to ten times more time to finish a job than they originally quoted. This additional time can drastically disrupt a time schedule for finishing a project, so if you're trying to stick to a schedule and you have to rely on a shop for something you are farming out, it may be a wise decision to start that work sooner rather than later. You can be performing all your parts sourcing, research, cleaning and refinishing parts, etc., while the car is at the body shop or while your engine is at the machine shop.

The Process

Restoring cars is a series of processes. Because every car is built differently, there are some differences in the order in which you perform each job to end up at the final destination. A complete build-up of a 1973 Ford Mustang may be different than restoring a 1967 Camaro, simply because these cars have different frame constructions. These types of different construction can drastically change the process of when and what parts need to be painted and assembled.

You're reading this book because you are restoring a first-generation Camaro, but the process would be the same for restoring a first-generation Firebird, because they are similar vehicles. The order of processes is suggested from the standpoint that some parts should be installed or painted before other parts; for example, the headliner should be installed before the windows are installed. Sure, you can install the headliner with the rear window and front windshield installed, but the process is much simpler if your car didn't have its front and back glass installed. (I'm not suggesting that you remove the windows simply to install a new headliner.)

The sequence of processes benefits the restorer to make life easier. Even if you're only performing two of the jobs on the list, you should still take the order in which they are written as a suggestion.

There are two different restoration methods when it comes to bodywork and paint. One approach is to perform all of the bodywork first (or at least the work that will affect the paint in the door and trunk jambs) and only paint the interior and trunk. Then paint the jambs (door and trunk jambs, window channels, firewall, panel edges to be bolted together, and any part that will be inaccessible once the body is assembled) the body color. After the body is completely assembled and the body panels are all aligned, paint the rest of the body and feather the edges to eliminate any harsh tape lines. Now you can almost build the entire car, with the exception of door handles, locks, trim, and any other parts covering an exterior painted body surface. Without the exterior body painted, you are free to perform assembly with less fear of damaging the new paint. The rest of the bodywork and paint can be performed, and the last pieces can be installed.

The second approach is to perform all the work and paint in a full-frame-off restoration process. In this process you completely disassemble the body shell and fenders before installing a single part, unless you need to temporarily assemble body parts for metallic paints or special striping. Separately paint all the parts inside and out and then carefully store the removable exterior body panels in protected blankets until a later time when they are installed. Carefully build your car making sure you don't damage anything during the long and tedious assembly process. If you do damage something, fix it when the car is completely done (if possible). The doors and truck lid were installed when GM painted the Camaros. When taking a Camaro apart, there is never paint behind the hinges.

Original or Reproduction?

Just about every restorer knows there's a difference between original factory and reproduction parts. As hard as aftermarket companies try to accurately manufacture reproduction parts, a lot of times something is lost in the translation. The differences range from the mounting studs on the taillight housing being a different diameter than the factory unit (rendering original nuts useless), to something as simple as the armrest covers on the interior door panels being soft and comfortable for your elbow, but the reproduction ones being as hard as a rock. Bolt holes in body panels may also not line up and require elongating the holes or drilling new ones.

There are different levels of quality when it comes to reproduction parts, and the old adage "you get what you pay for" certainly applies to reproduction parts for first-generation Camaros. Of course, some companies charge an arm and a leg for a low-quality part, so you have to watch yourself and do careful research to find the best parts for the money. However, in most cases, you pay a decent price for a decent reproduction part and you pay a lot more for an extremely good copy of an original part. Think of the difference between the quality of a screwdriver made by two tool companies, Snap-on and Allied. The screwdriver from both companies will do the same job, but the one from Snap-On is going to look better and work better for longer. It's not going to strip out your screws after a few uses and the handle isn't going to spin on the shaft when you apply a little torque.

The problem doesn't stop at the manufacturer. Some reproduction parts suppliers (not to be confused with parts manufacturers) carry parts of more levels of quality than other suppliers do. Good suppliers offer the same parts by two different manufacturers to give the restorer the choice of a cheap part and a higher-quality part. Some suppliers only carry the cheap part, and it's priced accordingly, but if you paid a few dollars more from a different supplier, you could get the better part. Keep that in mind when comparing prices of restoration parts.

You'll be surprised to know that some restoration parts suppliers don't even carry the best reproduction parts on the market, simply because the best parts are sometimes too expensive for the at-home restorer. They stock the parts they know the majority of us are willing to pay for. The bad thing is suppliers are making decisions about the quality of your final product without your knowledge. Again, educate yourself.

In rare instances, factory parts fit as well as or better than original parts because originals are better than factory parts. Body panels are the only exception to the rule. Dynacorn warns Camaro owners that NOS (New Old Stock) GM sheetmetal is not the same as original. Unknowingly, people get so excited about finding NOS GM sheetmetal. In fact, these parts are not typically of the same quality as originally used for assembling Camaros. They are replacement parts that were built after the production run of first-generation Camaros and could have been made from dies that were worn out. Many of the parts left out there on the market were sent to a dealership to be installed on a car and didn't fit, so other parts were sent. The part that didn't fit so well was shipped back to GM and set aside until all the good panels were gone. These inferior stampings were eventually all that was left to sell, and they sold for good money because buyers think they are getting the cream of the crop.

As a discriminating restorer, recognize that some GM-produced replacement parts are not always an exact replacement. For example, original GM factory replacement trunk panels had access holes on the underside for attaching Camaro and Firebird spoilers because the trunk fits both cars, but the original trunk lid for the Camaro only had access holes for the Camaro spoiler.

Aftermarket reproduction sheetmetal is offered in high and low quality, and you need to determine the level of quality before you buy it. I recommend investing in the highest quality parts you can afford. One reproduction supplier can sell reproduction sheetmetal from a couple different sources, which isn't a bad thing. One manufacturer may stamp a better floorpan than a quarter panel, and another manufacturer stamps a better quarter panel than a floorpan.

Although we don't have the space and means to identify every single high-quality body part for a car, there are several established and reputable sources for body parts, and we mention a few. You can always ask around if you have a question about the best source for a body panel. Usually the best sources on the Internet for constructive feedback are: www.yenko.net, www.camaros.net, www.pro-touring.com, www.lateralg.com, and other Camaro-based websites. Make sure you ask the other enthusiasts about their personal experiences with the panels, not about their opinions based on what they've heard or read.

Parts Sources

You can buy parts from many different sources. Some businesses, such as Year One and Classic Industries, strictly sell new parts for your Camaro. Then there are companies, such as Steve's Camaros, and D&R Classic Automotive, that base their business on selling new and used original parts. Other companies, such as GM Sports Salvage, do the majority of their business selling used parts off salvaged Camaros and other hard-to-find GM muscle cars, but also sell reproduction parts in their showroom. Muscle-car-based swap meets are also a good place to find reproduction and used original parts, and sometimes these are great places to find a good bargain on hard-to-find parts.

In the old days, I spent many weekend mornings scrounging hot rod swap meets in Turlock and Pleasanton, California, looking for little parts for my Camaro because those were the only real sources. This was back before the Internet, but these days, you could almost build an entire first-generation Camaro from all reproduction parts—and they can all be ordered over the Internet or by phone. There are still a few pieces not reproduced and there are also some parts that reproduction companies just can't match fitment and quality of an original part. That's when the parts sources selling good, used Camaro parts come in handy.

Documentation

Thoroughly documenting your car isn't a burden; it's a definite benefit, because you save time and prevent headaches by taking the guesswork out of reinstalling parts.

Before removing the first nut or bolt, get your camera out and start taking detailed shots of everything. Take a lot of pictures of small details such as bolt heads, weatherstrips, sealers, alignment gaps, body shims, suspension alignment shims, steering components, carpet, seats, pedal pads, wiring, number and letter stampings, and casting numbers (on body panels, axle housing, engine block, cylinder heads, carburetor, water pump, pulleys, transmission, shifter, radiator, engine compartment decals, window markings, etc.), brake parts, body gaps, visible rust damage, etc. Don't forget to take pictures of the damaged parts you may need to replace and missing parts so you can refer back to these later. Take pictures during the process of removing parts, so you know where all the parts go and which bolts and screws came out of what parts. There are many different-size trim screws for the interior and it helps in the long run if you have some sort of picture documentation on hand.

Thanks to the digital age, you can afford to take hundreds of photos. The images not only help when you are re-assembling your project, they also document the process of your project. A well-documented project also adds value to the restoration if you ever sell the car. Potential buyers typically pay more for a well-documented restoration.

Once you start disassembling the car and removing parts, put the fasteners in labeled baggies, tape them to the part and store the part, making sure to consider the tape may be in place for a long time depending on how long the project takes to complete. (Tape left on a part too long leaves adhesive residue behind that can possibly damage the finish of the part, or at the very least leave a messy sticky residue.) Don't forget to write "right" and "left" on parts removed from the left or right side of the car. This shouldn't need to be mentioned, but determining the left and the right side of the car is done by sitting in the driver's seat and looking over the hood.

If the part needs to be refinished, replaced, etc., make a note of it in a journal so you know what parts are needed to put the car back together. You could go as far as taking a picture of each part before you store it, just in case you need to e-mail some pictures to a friend or a shop. Keep the pages of the journal separate if you're doing a big job. Write a separate page for the interior, front suspension, rear suspension, steering, etc.

Organization

Most novice restorers start ripping every nut, bolt, and part off the car, and they put all the hardware in one pile and the parts in another. This creates a big mess because parts take up very little space when attached to the car, but take up a surprisingly large amount of space when removed from the car. Most parts don't store well due to size and shape. In fact, stored parts can take up almost the space of an entire car. Some of the hardest parts to store are fenders, fenderwells, seats, and interior door/side panels. Make sure you take this into consideration before removing parts. Have supplies on hand to make the organization a little easier. Large narrow boxes are good for storing door panels. Fenders and other front sheetmetal store well hanging on a sturdy wall with bicycle hooks or large nails.

Wire egg crates and heavy-duty plastic bins are great for large, heavy items, but smaller items tend to fall out. The downside is that these basket-type crates are not sealed and attract dirt, dust, and unwanted pests. Cardboard boxes are nice because parts won't fall out of them when they are closed. However, you can't see what's in the box. If you use them, make sure to label each box carefully, so you don't have to open 20 boxes to find a single part.

The best way to store parts is in closeable plastic storage bins. The lids close and lock together and can easily be stacked on shelves or in the corner of your garage. The heavy-duty bins are usually made of colored plastic, so label the outside of them. Light-duty bins are usually made of clear plastic, and it's a breeze to see what's inside when they are stacked in storage.

Buy some resealable plastic baggies in different sizes: snack, sandwich, and gallon. Get some permanent markers, some blue contractors tape, and some packing supplies. If you take a simple part like the door sill trim, put the screw in a snack bag, label the bag "sill plate screws," tape the bag to the sill plate, and store it. Do this for all the parts. It's more work now, but you'll be very glad you did it when you start putting your project back together.

Become a Pack Rat

Don't throw anything out! Keep everything, even your old curled-up door panels and other parts you don't think you'll ever use. If you don't keep everything, you inevitably run across little parts you can't get or parts you need for templates because the reproduction pieces don't fit properly. Don't sell your extra parts until you've finished the project or know for a fact you really won't need them. After you're done restoring your car, throw the part away or sell it to someone who can put it to better use. Your significant other won't be happy with all the old parts lying around, but they can pay dividends in the end. Store them as if they were still good parts. You may be surprised at the lack of quality of a reproduction part and decide fixing your original part would be a better option.

Disassembly

The best way to remove a windshield or back glass is to start by removing the trim. Use the correct windshield trim removal tool, which is available at most auto parts chain stores. Slide the hooked end of the tool under the edge of the clip, and pry the clip away from the trim, while giving a little lift to the trim at the same time. Compared to the aftermarket reproduction trim, the stock trim is a little more sturdy and thus forgiving when you pull up on it. Reproduction trim is made of softer material and is easily damaged. Replacement window trim is not cheap, especially original pieces, so I cannot stress enough to use the proper tool and carefully apply force.

Rear Window Trim Removal

Removal of the rear window is not difficult. Start at the lower left corner, and work your way in a counter-clockwise direction around the trim, only sliding the trim out of the adjoining trim when possible.

Windshield Trim Removal

The process of removing the windshield is much more involved. Start by removing the windshield wiper arms with the proper windshield wiper arm removal tool. Using any other tool may damage the paint on the cowl panel. Remove the cowl vent panel to access the lower trim attachment screws. There are two more trim clips holding the lower trim, which are located in the doorjamb under the front fenders. Remove the windshield's left-side

GETTING STARTED

trim, and work your way around the windshield using the same techniques as with removing the rear window trim.

Door Panel Removal

After you remove the door panels, carefully store and protect them, even if they are not in great shape. If you remove the door panels, store them flat or hang them as if they were attached to a door. Otherwise, they'll start to curl and the vinyl will start breaking loose from the board. You may find out that you can save old door panels if they aren't ripped up.

Weatherstrip Removal

Using a somewhat flexible plastic or Delrin blade-type tool you can remove old weatherstripping. Some of the weatherstrips are held in by glue, but some also have little plastic push-in retainers or screws helping them stay in place. You can pry the retainers out of the panels with a forked panel removal tool available from most specialty tool manufacturers or from Eastwood.

Windshield Installation

The factory (and proper) installation of a Camaro windshield is to use a tar-like strip of butyl. These days glass installers typically use a black urethane adhesive, which works great on new cars that have the dark-colored band of enamel baked into the perimeter of the glass. Our old windshields don't have this enamel band for the urethane to adhere to and the glass eventually breaks loose and causes major leaks. Some people swear by the urethane, but there's no substitute for the correct factory butyl ribbon seal for a long-lasting seal. The butyl is a sticky tar-like

This special hooked tool is required to safely remove the stainless window trim. This trim is very expensive to replace, so be very careful when removing it.

Stock pieces are a little stronger and more forgiving than aftermarket trim.

string, which is available in 5/16-inch and 3/8-inch diameters. The 5/16-inch diameter is the correct size if you're installing original glass, which is thicker than most aftermarket glass. Use 3/8-inch butyl if you're installing the thinner aftermarket glass. (Back glass is thinner than the windshield because it's not layered safety glass, so you would use the 3/8-inch butyl for the rear glass.) If you use the incorrect size you could have window trim fitment issues.

Any rust damage on the pinchweld or surrounding channel should be repaired, primered, and painted just like the outside body surfaces. Before going further, install both of the windshield glass supports at the base of the cowl to keep the windshield in place. (If you're installing a back glass, cut the rubber blocks that come in the package of butyl to the proper size.) Have a friend help you test fit the windshield without any sealer. Carefully lay the glass in place and find the proper center and height, which can be adjusted with the bottom glass supports.

Place a piece of tape on the glass above the lower glass supports and use a marker to indicate the proper location. This gives you a guide for installing the glass in the right location. Use this to your advantage because once you lay the glass on the butyl strip, you can't reposition the glass. It's a one-time deal. If you lift the glass, you'll mess up the seal or break the glass trying to pull it.

To confirm the pinchweld is ready for installing the windshield, use a good dirt/wax/grease-removing solvent to make sure the painted surface is ready to accept some pinchweld primer. Treat the pinchweld with the proper primer for butyl applications. This protects the pinchweld surface from corrosion and drastically promotes adhesion of the butyl to the pinchweld.

When installing the new window, keep in mind that the butyl can be seen from the outside of the car. The window trim doesn't cover it, so spend extra time being careful to make the butyl very straight. The butyl tends to stick to your fingers, so use a light touch. Start the butyl ribbon at one of the two lower corners of the pinchweld and carefully run the ribbon around the pinchweld, being sure not to pull it tight, which reduces the ribbon diameter and creates possible leaks. Make sure there's a good joint where the two ends meet.

CHAPTER 3

BODYWORK

The basic rule of thumb when working on cars is to only replace the parts that need to be replaced, with the exception of parts that come in sets, such as leaf springs, valve springs, brake pads, etc. If one quarter panel is rusted out or badly damaged, there's no reason to replace both unless they both need it. The same goes for floorpans and frame rails. The worst enemy of old muscle cars is rust. Sometimes it lurks under new body filler and recent paint jobs, so you don't know it exists until it's too late. It's common knowledge that once you see rust, there's a lot more you can't see yet.

What Should Be Replaced?

While restoring your Camaro, you'll often discover some sort of body damage that demands repair. After all, these Camaros are more than 40 years old, so there's a good chance that there's been some trauma—be it a serious dent or some rust—to at least one panel. With the help of your body shop and some knowledge shared in this and other publications, you can decide how to best deal with the damage. In most cases, you should keep as much of the original sheetmetal as possible, but you don't want to spend three times the cost of parts and labor to repair a stretched-out and rusted original door skin instead of replacing it. Don't get so emotionally attached to a factory piece of sheetmetal, or the entire shell for that matter, that your judgment is clouded and you keep parts of the car that just don't make sense.

It took about 40 hours of labor to get to this point. With all the parts welded together, it's time to add the window trim studs to the rear window gutter. Once you're sure all the welding is done, it's time to seal (not fill) the gaps between the quarters and the tail panel, and the gap at the rear window filler panel, with Kent Leak Check. If you do this with body filler or hardening sealer, it will eventually crack and allow moisture to enter the seam and start rusting. (Photo courtesy Steven Rupp)

BODYWORK

> ### Warning
>
> Some of the panels in the Camaro came galvanized from the factory. When you cut the galvanized steel with a plasma cutter or weld on it, the galvanizing burns and creates fumes, which are very hazardous to your health. Always make sure you are working in a well-ventilated area with plenty of fresh air. Some of the panels that are galvanized on first-generation Camaros are trunk lower side panels, rear frame rails, rocker panels, and leaf-spring brace panels.

If your car is an original collector car and it's going to make a difference that your deck lid isn't stamped with the same Fisher Body stamp as the trunk flange and other sheetmetal and you have money to restore the piece, it makes sense. But you will have to pay a very talented metal worker who really knows his way around hammers, dollies, and heat-shrinking techniques with heat sources, like an oxyacetylene torch or a resistance spot welder. Sometimes it makes more economical sense to replace a piece of sheetmetal with a new stamped panel, as long as you buy a good stamped part from a good source. Companies, such as Dynacorn and OER, produce panels that are made slightly thicker and with better metal than the factory used.

Common Rust Areas

Trouble areas for rust include the top of the dashboard where it meets the windshield, underside of the cowl (which can be seen by peaking your head under the dash), lower corners around the rear window, lower quarter panels in front or behind the rear tires, rear corners front fenders where they meet the rocker panels, floorpans, trunk pan, window rain gutters, etc. These are the first places Camaros rust. In fact, if you don't see rust in the common rust areas, there are two possibilities: The car is virtually rust free and there shouldn't be rust anywhere else on the car, or you can't find the common rust because somebody already repaired it.

However, there are rust areas that you can't see without more in-depth inspection or in places you can't see without removing trim parts or body panels. If somebody repaired the more visible areas and you find more hidden rust, you're in for a big surprise because they probably dug into a panel to repair it and found the damage was too extensive to continue, so they made the smaller

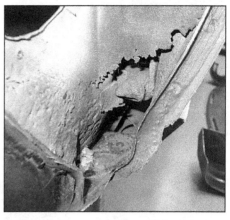

More severe rust, such as this area in the cowl drainage pocket behind the side kick panels, can fester in areas you cannot see without removing parts. This area traps dirt and leaves, which breeds rust because it also traps moisture. Rust in these places doesn't seem bad, but, with this structure rusted away, the front structure loses a lot of necessary strength. The outer panel has been cut away. (Photo courtesy Detroit Speed)

If you look at the underside of the dashboard, you may be surprised to see rust in places you never imagined. The factory did not apply much paint to these surfaces and rust is common. This Frankenstein-looking repair to the cowl water channel was removed from Stuart Adams' 1969 Camaro. (Photo courtesy Detroit Speed)

HOW TO RESTORE YOUR CAMARO 1967–1969 29

CHAPTER 3

Firewall Sheetmetal Installation

1 A Detroit Speed employee removes excess steel in preparation for installing all-new sheetmetal. Note the braces in the door openings and across the A pillar to keep the structure square, while all other support is missing. Also notice that the exterior paint, glass, and headliner are covered. A detail many shops miss is protecting these delicate surfaces from the hot flack when grinding and welding metal. (Photo courtesy Detroit Speed)

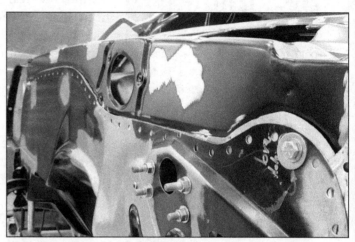

2 The panels are test fitted (note the brake-pedal bracket bolted to the firewall to ensure correct fitment) before being welded. They have all paint removed between the panels to promote clean welds, and have been drilled every couple of inches to be plug welded for strength. When the panels are three or more layers thick, all but the bottom panel are drilled, so all panels get good weld penetration. (Photo courtesy Detroit Speed)

3 All the holes in the firewall and upper and lower cowl are plug welded. This car is not a stock build, so there isn't a hole for the heater or blower motor. You can see the jig (bottom of photo) used to make sure the firewall subframe mounts stay lined up correctly during the process. (Photo courtesy Detroit Speed)

4 Make sure each panel is clamped across its entire length to keep it from shifting. Don't put too much heat in one area, to avoid warping. Weld the hole, cool the area, and start over. You can see that the plug welds on this firewall have all been ground smooth, starting with 3M Green Corps Roloc discs on a high-speed die grinder. The seam was then tig welded with Silicon Bronze filler rod. The strength of tigging with bronze is about 50 percent actual steel, so it needs to be accompanied by the plug welds. Silicon Bronze does not require much heat to tig so it's great to fill seams without warping the panels when you're careful. Clamps can also be seen inside the car on the dashboard. (Photo courtesy Detroit Speed)

BODYWORK

5 *The outer shell of the cowl panel shoulder is the last piece of the puzzle and has been prepped the same as the other panels. It is clamped in multiple places to ensure it stays where it's supposed to. Once the panel is welded in place you should paint the inside of the panel and seam seal it, since there's a good chance you'll still have access to the back side of this area through the fresh-air vent holes. (Photo courtesy Detroit Speed)*

6 *Once the welding is done to the internal seams, Detroit Speed primers all the surfaces with rust-inhibiting paint to protect the steel, then seals all the seams with SEM seam sealer to keep moisture from getting between the panels where rust breeds. Unlike the factory, you can seal the back side of the seams to protect your investment of time and money. The stock panel has a duct opening here, but a custom speaker can was built in its place. Once the outer shell is welded in place, painting and sealing access is very minimal, so it was sealed as much as possible, but the flanges weren't painted so they could be welded. (Photo courtesy Detroit Speed)*

repair and put the car up for sale. You may also find rust in the doorjambs between the body and the door hinges. Because the factory attached the hinges to the body and doors, as well as the deck lid to the trunk hinges, before paint or primer was applied, these bare-metal surfaces were prone to rust in extra-damp climates.

Just because you don't see rust in the common areas doesn't mean that there isn't rust in many other hidden areas.

Rustproof Coatings

You can cut rust out and replace the metal. Or, if the rust isn't substantial yet, you can use a good rust-inhibitor coating. There are a few great products on the market that stop oxygen from getting to the rust. Without oxygen, rust can't continue to spread. The two most popular products on the market are KBS Coatings and POR-15.

Floorpan Replacement

Because the external body panels are extremely visible, they get a lot more attention and repair. The floorpans typically are left to rot, or if they are repaired, they are hacked together and left alone. If you looked under a lot of Camaros, like I have while working at a shop with hundreds of Camaros up on mechanics' lifts, you'd be appalled at the quality of work (or lack thereof) performed on floorboards. I've seen some rusted-out floorpans where new patches were installed with screws and gasket sealer and some floor patch panels welded directly over the original rusty panel, which can still be seen from under the car. Hackers know that the carpet covers the top side of their hack jobs and a couple of cans of rubberized undercoating covers the work under the car. It's not easy to do a great job under the car, but it's not impossible.

Replacement floorpans come in different types. You can buy smaller patch panels in quarter sections, left half or right half, from the firewall to the rear seat or from the firewall to the tail panel.

Inspect your floorpan to determine how much you need to replace. If your floorpan has minimal rust on one side or in one part of the footwell, you may only need to replace a patch panel. Keep in mind that heavy surface rust can weaken a panel severely. If there aren't any visible cancer holes in the floor, it doesn't mean that the floor doesn't need to be replaced. If you can move the floorpan by simply pushing down on it with your hand, it indicates serious structural weakness and you should consider replacing the panel. The seats are bolted to the seat risers, which are attached to the floorpan. If the floorpan is weak enough and your seat breaks loose while you're driving, you could get in a serious accident.

When removing the bad sections of metal, keep removing material until the metal is not heavily rusted with scale. Replacement floorpans need to be welded to the old metal. It's hard to weld a new floorpan to rusted and thin sheetmetal. If you're successful in welding a new panel to a severely rust-damaged panel, the strength of the car is questionable. A structure is only as strong as its weakest link, and you want to perform the highest quality restoration possible, so the car provides years of reliable service.

Dynacorn sells replacement floorpans in a couple of different configurations. For severely damaged Camaros, it sells a floorpan that is completely welded from the base of the firewall to the tail panel. It includes a seamless floor section, welded trunk pan, and welded rear frame rails. The extra freight to ship this section outweighs the labor it would cost a shop to weld all the panels together.

Sound-Deadener Removal

With the introduction of sound deadeners that stick to the floor, such as Dynamat or Thermotec CoolMat, there's a whole new step added to the equation of restoration. Removing the sticky mat is almost impossible without making a mess, unless you use dry ice to freeze it. (Don't handle dry ice with your hands because you can get severe burn injuries similar to burns from heat and have to visit the emergency room at the hospital. Handle the sheets of dry ice with tongs and use insulated leather gloves.) Always wear protective eye goggles when handling and working around dry ice. The insulating mat is typically a gooey, sticky tar-like substance that takes considerable time to scrape and scrub off the panels. After most of it is removed you still have a sticky residue that takes hours to clean with lacquer thinner or other solvents. When you lay the dry ice on the insulating mat for a very short period of time (test on a small section to determine how long it takes for your specific application), it freezes the mat to a solid substance that is easier to chip off the panel.

Use a hammer and a heavy-duty putty knife to knock the sound deadener off the panel. Get the dry ice in flat square slabs that you can lay flat on the floor.

Quarter Panels

The market for the first-generation Camaro sheetmetal was not great in the late 1990s. When it came to replacement quarter panels, you had to get an old NOS Chevrolet quarter if you wanted one with the integrated sail panel. This was before eBay.com and Craigslist.com where you can just get online and have unlimited possibilities at your fingertips. If you didn't have access to knowledge of an NOS quarter panel stuffed in some guy's garage, you were forced to buy just the quarter skin without the sail panel or trunk sill, or you had to find a donor car with a good quarter. A lot of cars that we would call restorable today were cut up for quarters, then scrapped.

A few companies, such as OER and Dynacorn, have stepped up and met the demand because donor cars are more difficult to find and the steel is losing its integrity because of age.

Quarter Panel Replacement

1 *Overall this quarter panel was in pretty good shape. The person who replaced it many years ago performed a decent job of filling the gaps with body filler, but installed it about 1/4 inch too low. Dick Kvamme, of Best of Show Coachworks (BOS), is removing the bulk of the panel with a plasma cutter. He left the roof seam, doorjamb, and edges to be removed more carefully, then trim the rest with a cutoff wheel after cutting the spot welds loose. Be extra careful not to cut through the braces immediately under the panel in the sail panel, and the 1967 diagonal brace from the rocker panel to the doorjamb. (Photo courtesy Steven Rupp)*

Quarter Panel Replacement CONTINUED

2 *The previous installer forgot to reconnect the quarter to the outer wheelhouse as well as the rear package shelf to the inner sail panel support structure. The lack of integrity explains the cracked body filler in the quarter panel. This is poor quality from an inexperienced installer. (Photo courtesy Steven Rupp)*

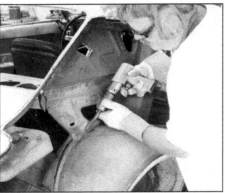

3 *Since the inner and outer wheelhouses had also been installed poorly, they were removed to make way for new stamped panels. (Photo courtesy Steven Rupp)*

4 *Don't be tempted to do the job halfway and not replace the outer doorjamb panel. The more splices you make the more places there are for future cracks to appear. Dick has drilled out the perimeter spot welds and carefully ground out the spot welds around the striker. The previous installer had put a 1967 quarter onto this 1968. You can tell because of the absence of the air vent in the doorjamb. With some attention, we were able to salvage the jamb sub-panel. We cut the vent hole after the job was finished. (Photo courtesy Steven Rupp)*

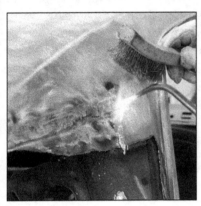

5 *The original roof seam was still in place because the quarter was only replaced up to the base of the sail panel. The factory seams are "leaded." Lead is bad for you and the environment, so take the proper safety steps to protect your skin and lungs when heating the lead and scrubbing it with a wire brush. Don't overheat the surrounding metal or you'll warp the roof panel. Only apply enough heat to the area to remove the lead. (Photo courtesy Steven Rupp)*

6 *Unfortunately, more harm than good was previously done on the inner wheelhouse and tail panel, so they were also removed. All the excess metal was trimmed around the edges with 3M cutting and grinding discs. Before removing the quarter and other pieces, Dick tacked in the angle iron to keep everything lined up. The whole trunk pan and rear frame rail will sag too, so it's important to support it from the bottom and make sure it's square before welding any of the panels in place. Here you can see a new trunk floor drop-off panel tacked into place. (Photo courtesy Steven Rupp)*

CHAPTER 3

Quarter Panel Replacement CONTINUED

7 The inner wheelhouses are very important structural supports for the rigidity of the rear section of the car. After multiple fitments of the inner and outer wheelhouse, it is finally welded in all the right spots. You can also see where the inner sail panel support is now welded to the rear package shelf. Dick confirmed the trunk was square, so he welded the trunk hinge support to the inner wheelhouse. (Photo courtesy Steven Rupp)

8 Once the outer wheelhouse is fitted and welded in place, it is time to protect all these hidden parts from inevitable future rust problems. We used some Rust Seal from KBS Coatings, but we didn't coat the areas we're going to weld. We poured the paint into a separate can, to not pollute the original can. (Photo courtesy Steven Rupp)

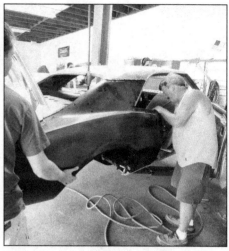

9 Gregg Blundell helped Dick test fit the quarter panel. It took many tries to confirm which edges on the panel and the car had to be trimmed before the panel fit the car like a glove. You can see the panel spreader giving the roof a little lift to help with the fitting of the panel. (Photo courtesy Steven Rupp)

10 It's common for the outer wheelhouse and the quarter to take extra massaging to align correctly. Even if all the panels are from the same supplier, this can be a problem area. It's just part of the job and it's expected. (Photo courtesy Steven Rupp)

11 Once the car is square, confirm that the trunk gap to the quarter and the doorjamb is good. Give the panel a couple tacks at the sail panel and the doorjamb. Before final welding is started, test fit the tail panel with some sheetmetal screws to keep it from moving around. The panel may take some trimming and adjustment to get it to fit to the quarters and the inner tail panel. (Photo courtesy Steven Rupp)

34 HOW TO RESTORE YOUR CAMARO 1967–1969

BODYWORK

12 With everything tacked in place, start welding the quarter and the roof panel together. The quality of the weld at the roof seam is very important because it is a high-stress area and you don't want the seam to flex. Be very careful not to overheat the roof or the quarter, because they warp easily, which adds even more time to your bodywork to fix. Dick welds a short section and then cools it with an air nozzle as he goes. (Photo courtesy Steven Rupp)

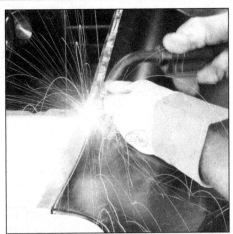

13 Drill a series of holes in the quarter panel gutter and plug weld them to the inner sail panel structure. Then place two very small tacks (that are removed later) to lock the gap between the quarter panel and the rear window filler panel while welding the rest of the perimeter of the panels to the car. (Photo courtesy Steven Rupp)

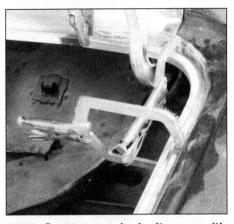

14 Quarter panels don't come with a trunk gutter that holds weatherstrip. The gutters usually come in a set of right, left, and upper center. Use multiple clamps to hold the gutter in place and don't get it too hot when welding. You don't want the gutter or the quarter to warp and distort. Small pieces like this become distorted really fast. (Photo courtesy Steven Rupp)

15 If you don't have the skill to "lead" the roof seam, use a filler that's extremely strong as the base filler. Once the weld is ground back, apply a layer of U-POL SMC Bonding Compound to the seam. It's a fiber-reinforced filler that works great for thicker applications such as this. Strength comes from the fibers, which includes carbon fiber. After only 20 minutes it was set up and ready for sanding. Dick wanted to knock it down in a hurry so he used a 40-grit 3M Hook-It sanding pad. (Photo courtesy Steven Rupp)

16 To make sure all the products work well together, use a thin coat of U-POL's Fly Weight lightweight body filler. After only 10 minutes it was set up and Dick "blocked" it with a short hand-block with 3M 40-grit paper to give the seam the correct shape. (Photo courtesy Steven Rupp)

Quarter Panel Replacement CONTINUED

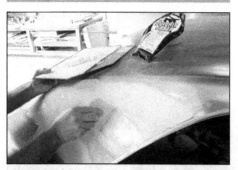

17 The final step involves applying a thin coat of high-viscosity U-POL Dolphin Glaze to fill any pinholes and imperfections. After about 10 minutes, use some 80-grit followed by some 120-grit sanding pads. Now it's time for some primer to prep the surface and get it ready for paint. (Photo courtesy Steven Rupp)

Body Panel Alignment

Just about every part of assembling a car has an order, which helps the process go smoothly. Aligning the doors and front sheetmetal is no exception. The subframe should be aligned to the body (see Chapter 9), so that the front sheetmetal that hangs off it can be installed correctly. Door alignment to the quarter panels and rockers is the best place to start. Once the door gaps are set, you can align the front sheetmetal correctly.

The radiator support should already be installed and the isolator bushings that attach it to the frame should be loose enough to allow adjustment during this process of aligning the body panels.

Fender Alignment

In order to properly align the front fenders, you need to complete the following procedures:

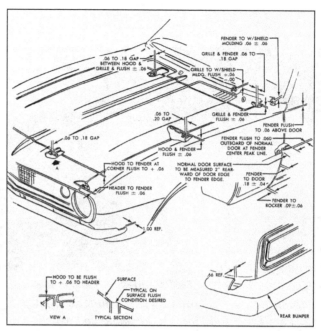

The front sheetmetal gaps are the same for all three years of the first-generation Camaro, with the exception of the front bumper measurements on the 1969, due to redesigned sheetmetal. Don't forget to adjust the doors to the main shell first. (This image is reprinted with the express consent of Year One Inc., a licensee of General Motors Service Operations)

Pre-assemble front fenders and skirts (inner fenderwells), corner bumper brackets, and fender extensions. Do not fully tighten the fender extensions; leave them a little loose for adjustment. If you're going to install new emblems on the front fenders between the front tire and the door, you should do that before putting the fenders on the car.

Assemble the headlight and all its hardware to the headlight housing bracket and set them aside for later use.

Assemble the lower valance, grille header panel, grille, hood catch assembly, and hood lock catch support.

Tape the edge of the doors and door gaps on the front fenders to add a little protection against chipping the paint during the process. This does not stop chipping from occurring, so you have to be careful.

Install all hood rubber bumpers (the two on the top of the radiator support and the two bumpers on the fenders in the hood gap).

Install the lower windshield trim on the cowl.

With doors closed, loosely install the bolts that hold the fenders to the radiator support. Loosely install the fender bolts on top of the cowl and in front of the cowl, as well as the skirt-to-firewall brace. Then loosely install the cowl vent grille panel at the base of the windshield.

Install head light and parking light wiring harness on the radiator support, then loosely install the header, grille, fender valance assembly, and radiator support.

Adjust the fender to the door and the cowl vent grille by shimming the top of the cowl and the front of the firewall. Tab-in the upper doorjamb, and then the lower fender bolts. Tighten the skirt-to-firewall brace.

Tighten the front sheetmetal bolts (including the front-of-fender skirts to the base of the radiator support), but leave the radiator support loose on the frame.

Using a tape measure, measure from corner to corner in an X pattern and square the front sheetmetal. Tighten the radiator support to the frame.

BODYWORK

Install the hood hinges, hood, and hood lock. Installing the hood itself is a two-person job.

Slowly lower the hood, making sure it clears the cowl vent grille cover and sides of fenders. Adjust the hood as needed to get the hood gaps to the fenders to match the gap to the front grille header, as well as the height of the hood to the rest of the body. Adjusting the hood is a difficult, time-consuming task.

Move the cowl vent grille forward to get the correct gap to the back of the hood, and shim the front as necessary. If the cowl vent grille moves too far forward, you may need to move the hood back a little to have the gap match the front and rear of the hood. If you have trouble with the gaps, you may need to loosen some of the body parts and shift them around a little.

Install headlight assemblies, grille parts, horns, and battery tray.

Dings and Dents

There are plenty of books available to show you how to repair dents and dings so I won't go into too much depth here. In fact, CarTech's *How to Paint Your Car on a Budget* is a great source. Eastwood sells a manual, *The Key to Metal Bumping*, which explains important dent removal basics and the right hammers to use for a particular job.

There are many different types of body hammers and they each perform a different task, so educate yourself on dent removal and figure out which tools you'll need. Eastwood offers a bodywork starter set that contains three different hammers with the most common profiles, so you can effectively work out dents and other body deformities.

Once you've done that, you can get good-quality tools from companies such as Eastwood. Don't forget, you get what you pay for. Cheap hammers are made with cheap wood handles that crack and break, and the metal faces and tips are soft, leaving you with junk tools after very little use. Even Eastwood's least-expensive hammers are better than the substandard hammers you can buy for a fraction of the price at the flea market or discount tool store.

One basic idea is that the best way to pull out a dent is to pull it out in the direction it was made. For instance, if you have a Camaro that was smashed in the rear and you took it to a frame shop, they would load the car on a frame machine. They would mount frame clamps and plates to the rear of the body and pull the deepest caved-in area out in the direction it was created. They would pull the dent out a little at a time to make sure they didn't pull it out too far. If the chains are not anchored to the car on the sturdy sheetmetal with good hardware you can pull the anchor right out of the panel, making the damage worse.

The best-fitting front fenders you can get are originals, and the next best ones are the OER fenders from Classic Industries, which use the original GM tooling. These cost a lot more that other fenders, but aligning the panels is much easier with good stampings. If you're starting from scratch, align the doors to the rockers and quarter panels. Temporarily tape the door and door gap on the fenders to help prevent paint chips—be extremely careful. Hang the front of the fender on the radiator support and hand thread in both bolts. Loosely shim the fender from the top of the cowl until the top body line matches the door while minding the gaps. Shim the fender on the front of the firewall, then do the same on the bottom of the fender by the rocker and the lower front bolt. If your shims are old and rusty, replate them or buy new ones. Otherwise, rust starts breeding right away. (Photo courtesy Frank Arone and Brian Henderson)

CHAPTER 3

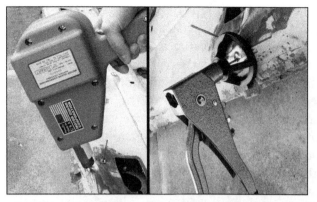

Eastwood sells this awesome stud welding gun. Spot weld the stud to the panel and pull it with the supplied slide hammer or optional handheld puller. These tools work great with about 15 minutes of work on a practice panel. Use a cut-off wheel to grind off the stud and you're ready for minimal body filler. No more big stripped-out holes to screw a slide hammer into a panel.

Other dents in the tail panel should be pounded out with body hammers, while tension is still pulling on the panel. Once the tail panel is pulled out enough, smaller bodywork can be performed. I've watched a very experienced body man work on a car that had rear-end damage and was hit hard enough to slightly buckle the quarter panel. The body technician anchored one end of a chain to the rear subframe and the other end of the chain to a sturdy anchor in the floor. He slowly pulled the car forward to gently put tension on the chain, then he gently got on the throttle. The tires spun for a few revolutions and the buckle in the quarter disappeared. A professional did this in a safe environment, so if you're going to try this sort of thing, use good judgment, an extremely strong chain, and a really good anchor on both ends.

The same procedure for pulling out a caved-in tail panel should be applied to smaller dents. In the past, the typical shop removed smaller dents with a hammer. They'd resort to drilling holes in the dent and would use a slide hammer to pull the dent out. Not only does this require drilling holes in the panel you're trying to repair, the screw of the hammer stretches the metal and pulls out of the panel, which causes even more damage to be repaired. In addition, the drilled holes also weaken the panel. But, fortunately, slide hammering has come out of the dark ages. Now you can tack weld rods to the panel and then attach the slide hammer to the rods and pull the dent out. Next, you simply snap off the rods and grind their heads off the panel. The panel is still in great shape and ready for some filler.

Paint Removal

You wouldn't knock down an old Victorian house and build a skyscraper on the old foundation. The same applies to paint. In order to get the best results on the body and paint, you're better off starting with a clean slate. You have no idea what is under that old paint or what condition the metal is in. Sure, you can use the old paint as a base and only sand the necessary parts down to the metal. Collision shops have been doing this for years with hit-and-miss results. Leaving the old paint saves time and keeps the cost down on a project, but it's a certain amount of laziness. I'm just saying that if you want the best results from your work, you should start with a clean slate. If you're going to apply body filler, you definitely need to remove all the paint first.

You don't have to remove all the paint if you can verify the quality of work and materials used so far in the project. If you've partially painted the car and determined its quality, you can match the paint. For example, the project car in this book had a few body panels and other parts painted a couple of years ago. We know exactly what materials were used and that all the work was done correctly. Now the choice has been made to change the color. In this case, we can simply scuff the current paint and apply new color, knowing

Applying chemical stripper is one method of removing paint and filler from the body. Make sure you completely remove all the chemicals from the nooks, crannies, and cracks. Neutralize the chemicals before applying primer to avoid a huge mess. (Photo Courtesy Detroit Speed)

with confidence that the foundation is top-notch. Ninety-nine percent of the time, you have no idea what's been done before, so removing everything is the safest way to proceed before you spend a lot of money and time. Removing years of paint, primer, and filler requires aggressive methods. Each way has good and bad points to get to the final result.

Sanding

Physically grinding and sanding the paint and filler from your car down to the steel is a huge task. Not only is this method extremely time consuming, it costs a lot of money in sandpaper. You can use a right-angle grinder/sander, which doubles as the polisher, to remove paint with a little more speed. These are very aggressive and can leave deep scratches in the metal, removing precious material. They also distort panels from the heat they generate on the surface of the panel. If you only use an orbital DA sander with less aggressive paper, you'll be spending a lot of time removing material and you'll spend a lot of money in paper, especially if your car has a lot of filler to be removed.

Chemical Stripping

There are two ways to chemically strip a car: You can either brush on chemical stripper or you may find a shop that can dip your entire car and panels in a large tank of paint-removing caustic chemicals. Applying chemical stripper by hand and then scraping and scrubbing off paint is a tedious and time-consuming job. Certainly, acid dipping the body at a body shop is far easier and less time-consuming, but it is also more expensive. If you choose to apply chemical stripper by hand, there are many stripping solutions on the market, including those that contain methyl chloride, which is an extremely hazardous substance. There are environmentally safe and effective options such as Eastwood's DeKote, which doesn't contain methyl chloride. DeKote effectively cuts through paint and makes scraping off paint relatively easy, and it's not potentially hazardous to your health.

Media Blasting

In the old days, you only had a regular sand blaster to remove all the paint from your car. The sand does a pretty decent job of removing paint, rust, and body filler from body panels, but if you're not careful, it also does a great job damaging the sheetmetal. The typical media used in sand blasting heats up the panels extremely fast, which causes the metal to warp and distort. A technique of keeping the blast nozzle at a 45-degree angle to the surface of the metal helps slow the damaging heat, but you should keep checking the surface to ensure you don't overheat it. The good part of sand blasting is that the bare metal surface is left with a rough texture, which promotes filler and primer adhesion.

The newer technique of baking-soda blasting is a much more elegant

Eastwood makes this affordable baking-soda blaster. The unit comes as a single-hopper that is filled with either soda or abrasive media. Eastwood also offers a dual hopper so you can switch between the types of media on the fly to suit your needs. (Photo Courtesy Eastwood)

way to remove paint and thinner body fillers. The baking-soda blaster can delicately remove paint by the layer. You can remove the paint from an aluminum can without damaging the can. The beauty of baking-soda blasting is that it doesn't damage glass, rubber, or chrome, so you don't have to do as much time-consuming taping and protecting of these surfaces. Also, the mess left behind is not abrasive like sand and other material, so you can also use it to clean engine compartments. You should still cover openings in the engine, but sand and other abrasive media is more harmful to moving parts than errant baking-soda dust.

When you're done making a mess on the ground, you have baking soda and the materials you removed from the car, instead of hazardous chemical strippers and media that can be harmful to the environment. Because you're probably removing grease, oil, and some lead-based products, you should wear a dust mask or respirator and eye goggles to protect your lungs and eyes. Wear a long-sleeved shirt to cover your arms, and leave the shirt untucked and over the top of your pants pockets to keep the debris and dust from building up in your pants. The mess on the ground should be swept up and disposed of properly because it contains some harmful debris that will pollute the soil and water supply.

Body Stamp Care

If you're doing a high-quality restoration, you'll probably want to be able to read the body stamps after the car is painted. Before spraying thick filler primers/sandable primers, take note of all the body stamps and tape over them before spraying the first coat of thick primer so that you don't fill up all the body stamps because they'll be unreadable. This especially goes for VIN plates and cowl tags (a.k.a. trim tags). We've personally seen VIN plates with so much filler primer on them that you couldn't make out 90 percent of the digits. The tops of the dashboards on these cars were obviously wearing a thick coat of primer hiding some ugly repair work. Imagine the first time a police officer notices the VIN is unreadable. Your day will get long really quick. We've also heard of a cowl tag filled with so much filler primer that when it was removed to clean it, there was a 1/16-inch ridge around the edges where the tag was sunken into the filler.

Once all the heavy primer has been used in the process of getting a straight body, peel off the tape, and carefully sand down the built-up edges of primer to feather them. Now spray your final primer, being careful not to put too much of it on the numbers. Then you are ready for paint, and the body stamps are much easier to read.

Cowl Tag Removal

Most enthusiasts have a big problem with removing the cowl tag from the firewall for fear of being fraudulent. There's nothing fraudulent about removing the cowl tag during your restoration as long as you put the same tag back on the same car. Tampering with the VIN plate is a different story.

You may need to remove the cowl tag for a few reasons, including to protect it from getting too much filler primer on it, from getting damaged during chemical stripping (remember that the tag is aluminum not steel), or from getting damaged during the build process. In addition, there might be rust behind the panel that needs to be treated and the cowl panel needs to be replaced, etc. Before removing the tag, take a good clear photo of the tag, just in case you lose it—with a two- to five-year restoration, that happens more than you would think.

Drilling out the rivets is not the best way to remove the plate because the rivets inevitably loosen and spin for eternity. With the cowl vent cover removed at the base of the windshield and the wiper arms and wiper motor removed, you should be able to squeeze a hand and some pliers behind the cowl tag. Gently pinch the back side of the rivets around the whole outside

If you're going to shoot a bunch of thick filler primer on the car, tape over the body stamps so they don't get filled and become unreadable. Pull the tape, feather the edges of the filler primer, and shoot your final primer.

of the rivet to make it smaller in diameter. Spend a few patient minutes doing this and the rivet should get small enough to push through from the back side.

Put the cowl tag in a special and safe place. Throwing it in your toolbox is a good way to damage it in a hurry because a busy toolbox has a lot of tools rolling around. If you're going to do that, you may as well just throw the tag out in the street. You can print out a picture of your cowl tag and put that in your toolbox if you want to. When you've done all the work and you're ready for a final coat of primer followed by paint, then it's time to install the cowl tag. You can buy factory-looking rivets from TrimTags.com and probably a couple other sources. They also sell a rivet install tool for a good price, or you can simply put a little dab of epoxy in the firewall holes, insert the rivets, and expand the rivets from the back side in the cowl area.

Once the rivets are holding the cowl tag in place, put a dab of seam sealer in the center of the rivet like the factory did. Now you're ready to apply a thin coat of primer and start painting the car and firewall.

The rust behind this trim tag obviously isn't sleeping. Remove the cowl/trim tag, treat the rust, and reinstall the trim tag with rivets from TrimTags.com. Notice the plant code is VN (Van Nuys) instead of LOS as it was earlier in 1969. (Photo courtesy Dennis Doherty)

CHAPTER 4

THE PAINTING PROCESS

There are many books on the market to teach you how to paint your car, and I have one chapter to devote to the subject. With that being the case, I cover the general principles, products, and techniques, and I also highlight items of specific importance to first-generation Camaros.

Most respected painters will tell you that it's possible to paint your car in your garage at home and get great results. But it is very challenging to do so. To achieve professional-grade results, you must build a temporary paint booth, so you can create a very clean environment. The key to great results is great bodywork, good prep work, keeping debris out of the basecoat and clearcoat, and knowing how to fix inevitable defects. In addition you need to be adept at color sanding, cutting, and polishing. You don't need a professional paint booth to get great results.

Can You Paint?

With the right tools and some practice, anyone can lay some paint on a panel. Spraying paint well doesn't mean anything if you didn't perform a good job on the bodywork and prep. Bodywork and preparation steps are 70 percent of a professional-caliber paint job. Color sanding and polishing after painting is 20 percent of that same paint job. That leaves 10 percent for the quality of a paint job right in between the two extremely tedious tasks before and after the paint is laid down. Some people just don't have the patience to spend the time or give the attention to detail it takes to do a great job. That's one reason why paint shops charge so much money. If you have more time and patience than you have money, you can hone your skills on bodywork and painting to save yourself a lot of dough. Take a few minutes to do some serious introspection so you

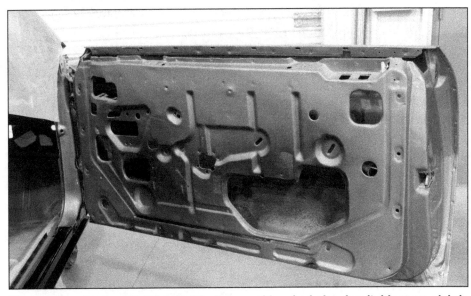

Blow all the debris out of the door before masking the holes. Inevitably some debris will be left in the doors and will blow out as soon as you start spraying paint. Don't mask the window opening at the top of the door. You need to get a good coat of paint in there. (Photo courtesy Brian Henderson)

HOW TO RESTORE YOUR CAMARO 1967–1969

can make an honest judgment of whether you will be hurting or helping yourself by performing your bodywork and/or painting.

Attention to detail and determination will take you far, but only if you also own or have access to the right tools. You need a good air compressor equipped with an adjustable regulator and a good dryer to effectively power the necessary DA sander, paint gun, and other tools.

Why are Shops so Expensive?

Bodywork is a time-consuming job, and you're going to pay good money for experienced employees to perform good work, but that's not the only reason body shops are so expensive. Painting a car is expensive, and it keeps getting more expensive. Two-stage paint (basecoat and clearcoat) requires buying both parts, where single-stage paints in the past meant that you bought the color paint you wanted and that was it. Paints are getting more expensive all the time because environmentalists and government agencies are constantly pushing new legislation to make paints more environmentally friendly. Every time paint manufacturers have to make "safer" paints, whether it helps or hurts the quality, integrity, or life span of the paint, the costs increase. In an effort to keep the environment clean, shops in most areas are required by law to only spray specific types of paint and use expensive equipment to filter what gets pumped into the atmosphere.

What Can Go Wrong?

Before you jump in with both feet, you need to know a few important facts about painting and bodywork and what can go wrong. There are some common misconceptions and mistakes that paint books and magazine articles don't stress enough.

Paint and primer are not scratch filler, so if you can feel or see defects or scratches with the naked eye, they will be ten times as visible once the paint is on the car. If a panel is slightly bent or your body filler is not sculpted just right, it will stand out like a sore thumb. This also includes crazed or excessively cracked paint. (Crazed paint is caused by paint being too thick or paint going over a coat that had not fully cured or adhered to the panel.) If your paint is crazed, you can't put enough paint

Caution!

Spraying or working with coatings, primers, paints, solvents, thinners, and other chemicals poses a number of health risks that should be written on labels or expressed with other accompanying information. Follow the directions as if your life depends on it, because in most cases it does. The chemicals and fumes are not the only a breathing danger, the fumes are also very flammable, so make sure you're always using them in a very open area with plenty of ventilation. Also make sure you're not working around an open flame, i.e., water heater, natural gas clothes dryer, furnace, etc. In the old days, when spraying lacquer paints, you could get away with using a good dust mask to breathe safely because they were non-toxic. Nowadays, the new paints are extremely toxic and require a filtered respirator to keep your lungs safe.

on it to hide it. If your old paint is crazed, you must remove the paint all the way down to the metal and start over. If you fill over the cracks with a specialized filling primer and it comes out as smooth as glass, those cracks will be back in about a month. If you have any defects in paint that aren't bugs or dirt, you've got to take the paint down to the metal and start over or they will be back.

You can't just sand down rust "a little" and then primer and paint over the top of it because it does not eliminate the underlying rust problem. Improper rust repair returns with a vengeance. Body filler and paint shrinks over time. You can fill a dent and have the filler perfectly sculpted and paint over it. After a few weeks or months go by, you'd be surprised at how bad the panel can look. The quality of the products used and their proper application cuts down on the shrinkage, but it's an important issue to consider. You need to perform as much work to the metal and use as little filler as possible to keep the shrinkage to a minimum. Less filler means less shrinkage.

Don't use WD-40 or another oil-based lubricant in or around primered or filled panels. Oils can become trapped in fillers and wreak havoc on paint. If you see fisheyes or any other chemically caused defects show up in a small section when spraying your paint, you can't add enough paint to cover them. If it's a small area, finish painting the car and repair that area later. If it's a large area and you just started painting, you should simply stop and thoroughly clean off the wet paint with solvent and a rag, but make sure you don't disturb the work under the paint.

Once the paint is removed, use solvent to wash the surface again, and remove the culprit causing the problem. I've heard of this happening a few times on the same area with stubborn silicone-spray contamination, where more intense solvents were used each time until one finally worked right. The next step would have been stripping to start with a clean slate. This makes a mess, but saves you a lot of sanding.

Compressed air is used to transfer the primer and paint to the surface of the car. When the air shoots out of the tip of the gun, it stirs up any loose debris it comes in contact with, so going over the surface of your project with an air nozzle is very important. Keep in mind that any loose tape or masking paper invariably causes you difficulty if you don't address it before you start shooting paint.

Decisions about the Outcome

What do you really want out of your car when it's done? The outcome of your paint job depends on how much time you spend on the bodywork, prep work, painting, color sanding, and buffing. The time spent directly reflects which part of the spectrum your paint job rests in. The spectrum starts at "good from afar but far from good," which usually looks good from about 10 feet away.

The next level is "looks good standing right next to it, but don't look too close." It's not too nice that you can't feel comfortable driving it every day and parking it just about anywhere without worrying about it.

Add about 40 hours of prep work and about 60 hours of color sanding and buffing to get to the next level, "nice enough that it stands out in a sea of other cars, where people are just plain mesmerized by how good it looks, but can't pinpoint why they can't take their eyes off it." This kind of paint job quality makes you nervous to park within 300 yards of the next car unless it's at a car show.

The highest level is "pure show." Add 40 more hours to each level of the prep, painting, and finishing, then double them again. The paint is so smooth that it appears to be a mile deep. This kind of car is kept indoors, usually in a car collection. You put 50 miles or less on this car per year. It's trailered in an enclosed trailer to and from car shows and when it is at a car show, you have to spend all day next to your car to keep people from touching it.

The outcome basically comes down to how much time (and money, if you're paying somebody else to do the work for you) you want to spend on making your car look good.

How Correct?

There's one more level that actually falls somewhere lower on the quality spectrum, but the owners of these paint jobs treat them as the highest level of quality. This paint is the factory paint. The factory performed a somewhat decent paint job for a production car of the time. The panels weren't awesome, the paint itself was not high-quality, it was thin on some sections of the body, and it was applied right over errant debris and seam sealer. Humans, using a stationary paint system, applied the paint. They were painting cars because they needed a job, not because they wanted to make a name for themselves in the artistry of the auto-painting world. The car bodies were painted in a hurry because it was an assembly-line process, and more

The factory sprayed a thick coat of semi-gloss black paint on the firewall. Original-paint cars have sags in the paint all over the firewall, but they are more prominent on the lower sections. (Photo courtesy Brian Henderson).

cars were waiting in line. To get paint like this you must spend a lot of time doing research on how the factory processed paint at the time your car was produced and you need to pay a very competent shop a lot of money to copy the mediocre factory paint.

If you are intrigued by the factory "correct" paint, I have included some information about the deficiencies. For instance, the paint on the firewall was laid on pretty thick and commonly had runs all over the face, but most commonly toward the bottom of the face of the firewall where the transmission tunnel starts to curve under the car. Seam sealer between the tail panel and the quarter panel (above the rear bumper) and in the rain gutters was never perfect.

A great paint job is 90-percent prep work and 10-percent painting the car. Any scuff marks or scratches won't fill in with paint. If there's an imperfection in the primer before you paint your car, it will be visible after the paint has been laid on the car. If you don't have the equipment to paint a car, you're going to need basic items such as paint gun, air compressor, dryer, and hose.

Start on small projects before you spend time painting your important parts. Pick up a junk hood or fender at a local discount auto wrecker, or find a rusted or damaged Camaro hood at a salvage yard. This way you can perform bodywork and test paint a hood that you can hang on your garage wall as art. Who knows? Maybe you'll do a nice job on the hood and have an extra hood you can be proud to bolt onto your Camaro when you're done. It's nice to have a part where you can hone your painting skills without having to make a mess of good parts.

Interior Paint

The Camaro interior has varying amounts of exposed metal, which needs to be painted. Chapter 12 includes some information about the painting, but the best time to paint these exposed metal surfaces is when you're painting the car. That's if you're painting the car with the windows removed and you're painting a bare shell. You won't have as nice of a finished product if you're painting the interior sheetmetal with the windows installed. I've painted dashboards with the windshield installed and with it out, and noticed a large difference in the quality. You just can't sand, prep, or paint the top of the dash panel adequately with the glass installed.

There's a 95-percent chance of some rust hidden in the window channel or on the top of the dash panel that needs to be attended to; so while you're at it, remove the windshield when painting the dashboard. The same goes for the back window, since any revealed rust should be addressed before the body goes in for final painting. The factory did not perform a great job of protecting metal with paint on the interior. It really dropped the ball when it came to painting the interior sheetmetal around the rear window. We assume it didn't think the area around the back window or any interior sheetmetal would ever come into contact with moisture. If you know anything about Camaros, you know it's common to have rust and water around the rear window sheetmetal.

For the rear package tray, the factory installed the metal corner trim covers over bare metal, then the dash and rear package tray area were painted the interior color. The interior sheetmetal above the door and

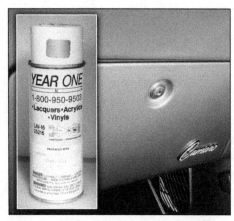

Year One and other companies offer original interior lacquer paint colors in 1-pint cans or 12-ounce spray cans. The correct factory finish is semi-gloss if you are looking for a stock look.

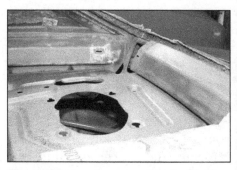

The 1968 panel was painted with blue interior trim color. This car was originally equipped with blue interior, Grotto Blue exterior, and white vinyl top. Notice the lack of factory paint with the corner trim removed. This area was bare steel when the trim piece was installed.

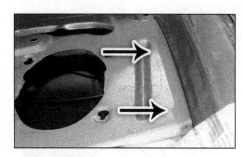

The same 1968 package tray shows where the factory masked off the interior paint, which was sprayed after initial primer coats. After sanding the primer and prepping the panels, the body went to the paint shop operations area.

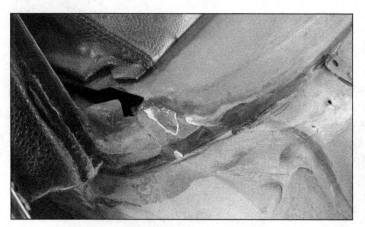

The top of the dash was masked off at the same time as the rear package tray area. You can see the tape line where the interior paint was masked before the body color was sprayed on.

THE PAINTING PROCESS

Depending on what you are looking for a finished product, you can paint the underside of your car all one color or you can paint it semi-gloss black and gray with body-colored overspray on the sides to replicate the factory finish. The 1967, 1968, and 1969 models all had different finishes. (Photo courtesy Brian Henderson)

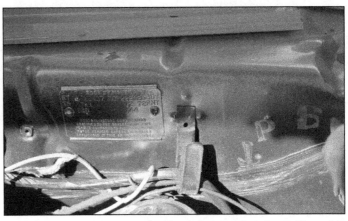

The 1968-69 Norwood (NOR) cars received the P, T, and B stamps on the left and right face of the firewall to confirm the body had passed the Paint, Trim, and Body inspections and was ready to be sent next door from Fisher Body to the Chevrolet part of the facility. Cars built at Los Angeles (LOS or VN) did not get these stamps. (Photo courtesy Brian Henderson)

quarter panels on the 1967 models were painted the interior color at the same time as the dash and package tray (unlike the 1968 and 1969, these door and quarter interior panels did not cover the top of the interior sheetmetal). The 1968 and 1969 interiors were painted like the 1967 interior, but the doors were not painted. Before the body went to paint, the interior paint was masked to protect it during the rest of the process.

The factory-correct finish on the interior sheetmetal and (only 1967) top of doors and quarter panels, and other trim (such as steering column and its trim, console, gauge cluster, standard-interior armrest bases (on 1968 and 1969), etc., is semi-gloss. The dash tops and rear package tray surround were painted more of a matte finish which, depending on the interior color, might have been a shade darker than the color of the rest of the paint. Restoration companies like Year One sell original-color lacquer paints (as well as dyes for fabrics) in 1-pint cans or 12-ounce spray cans.

Underbody Paint

The underbody was originally painted semi-gloss black before the exterior got its body color. The factory installed the six oval-shaped plugs in the floor and trunk with a white seam sealer after the floor was painted black, but before the body color was sprayed. During the body-painting process, the underbody and floor plugs ended up getting a good misting of the body color on the whole floor. Because it's not easy to reproduce the correct factory processes in the original order, you can mask the body and shoot the floorpan to your liking, and then mask the entire floorpan and paint the exterior. Mask the underbody off and mask the firewall up to the cowl drainage panel seam before shooting the body color. This keeps overspray on the firewall to a minimum.

After the car was painted, the factory sprayed a sound-reducing undercoating in the rear wheelhouses. It didn't perform a great job of keeping it in the fenderwells. Some coating overspray also made its way to some surrounding underbody sheetmetal, such as the side of the rear frame rail and quarter panel extensions. This undercoating also helped seal the center seam in the fenderwell. The undercoating applied at the factory never sealed the flange between the quarter panel wheelhouse and the outer wheelhouse. Moisture and dirt gets trapped in this area, and it is a common place for rust to start. If you have a chance, spend a few extra minutes sealing the flange all the way around the wheel opening where the two pieces of sheetmetal meet. If you're careful and do a good job, nobody would ever know you did it.

Firewall Paint

The factory painted the firewall semi-gloss black after the main body was painted. The amount of paint on the firewall depended on the factory,

CHAPTER 4

The factory semi-gloss black paint on the firewall does not reach the top edge of the cowl and does not have a defined tape line. Originally, the firewall was painted quickly on an assembly line after the body color was painted.

year, week, day, hour, and the worker who applied the firewall paint. Some firewalls were painted from the floorpan to the top edge of the firewall and some were painted only about 2 inches from the top edge of the firewall. In rare cases, the firewall didn't even get black paint, so it remained the body color. Some old photos of "press" cars (promotional vehicles for magazines) have the firewall painted over the top of the cowl, but this is not the norm. Each car was a little different because the factory worker simply sprayed the firewall by hand.

Dupli-Color has a spray paint that best matches the factory finish. Professional restorers prefer it because it's easy to control the factory finish with its runs and defects. Spraying the firewall with a spray gun is much harder to control the paint, especially the overspray so late in the project process. The paint is from Dupli-Color's Acrylic Enamel premium general-purpose paint line. The part number is DA 1603. It's the only paint sprayed from a can on Henderson's restorations, and the results are award winning.

Cowl Paint Differences

The factory painted the top of the cowl the same color as the body on first-generation Camaros. Therefore, a black RS came with the top of the cowl painted gloss black and a Rally Green 6-cylinder car came with the top of the cowl painted Rally Green. There were a few exceptions to this rule, which were the Z28 (if equipped with stripes), Z11 (Indy Pace Car convertible), and Z10 (coupe with Indy Pace Car paint scheme). These cars received an additional color on the top of the cowl.

In the case of the Z28s, the top of the cowl was painted the same color as the stripe. For instance, if the car was LeMans Blue with white Z28 stripes, the top of the cowl was painted white. Some have considered Z28s with white cowls as strange looking, and in some cases, original-paint cars judged at car shows have been marked as incorrectly restored even though the paint was original and therefore correct. Plenty of these white cowls have historically been painted over with black paint in an effort to fix what was thought of as a factory mistake or a detail simply overlooked during a restoration.

The entire cowl was not always painted stripe color. This Norwood-built Z11 (pace car) was only painted orange around the opening of the cowl, which is visible through the vent grilles on the cowl cover. This format is more common on NOR-built cars.

First-generation Z28s had either a small section or the entire top of the cowl painted the same color as their stripes. There's no rhyme or reason to entire or partial cowl painting. Contrary to popular belief, Original-Paint Los Angeles and Norwood-built survivor Z28s have been found with full-cowl paint. This 03D LOS 68 Z28's white cowl matches its stripes. Black-stripe Zs had gloss-black cowls. (Photo courtesy Richie Collins at Carbuffs)

Because these cars were built in Norwood, Ohio, and Los Angeles (Van Nuys), California, there are some differences between the size of the area painted the stripe color. Some cars had the stripe color from the base of the windshield to the top edge of the firewall and from the pinch weld flange on the left tulip panel to the flange on the right tulip panel, which are hidden under the tops of the front fenders. Some cars had just the area around the openings in the top of the cowl (which can be seen through the fresh-air openings in the cowl cover at the base of the windshield) painted the stripe color. Restorers have found both methods on original-paint Norwood cars and Los Angeles–built cars. The stripe-colored paint seems to be painted freehand sometimes and appears to have used some sort of template, but none have been found with a defined raised edge as if some sort of tape had been applied before the paint was sprayed.

If you're going to paint the cowl panel, keep in mind that the windshield wiper transmission arms inside the cowl were still painted semi-gloss black. So if you paint the cowl white, don't paint the arms if you're going through the trouble of restoring your car to factory-correct finishes.

Black Cowl Paint to Firewall Transition

If the Camaro is black, or it's a Z28 with black Z28 stripes and the full black cowl treatment, the paint on the top of the cowl should be gloss black. The body color was painted before the semi-gloss black was painted on the face of the firewall. The top of your cowl should be gloss black and the face of the firewall should be semi-gloss black. For factory-looking paint, don't paint the semi-gloss firewall up and over the top of the cowl. The top of the cowl was not painted semi-gloss black from the factory unless there was a huge mistake made at the factory while the workers were painting the firewall, which could have happened, but it's not likely. For a factory look, follow the guidelines in Firewall Paint on page 46.

The blower motor case (the heater core duct on the engine side of the firewall) was originally painted almost a gloss black from the factory, which contrasted with the semi-gloss black firewall. The case was one of the very few items in the engine compartment that was painted anything close to gloss black. The factory blower motor was painted semi-gloss black. A little tip: The original units have the build date stamped on their mounting flange, but replacement units do not. Over the years, these have a tendency to fail and be replaced without trying to save the original flanges. If you care to keep your car as original as possible, try to rebuild the original one with new parts.

Trunk Paint

From the factory, the trunk area received speckled-looking trunk paint. This paint resembles what most people refer to as "spatter paint." The trunk paint applied at Norwood was different than paint applied at Van Nuys. Norwood's paint is described as a dark base color

This is original Norwood trunk paint on a 1969 Camaro trunk hinge brace. Also notice the two different trunk spring rods. The lower (larger diameter) rod was installed on D80 (front and rear spoiler) equipped Camaros. (Tony Lucas Collection)

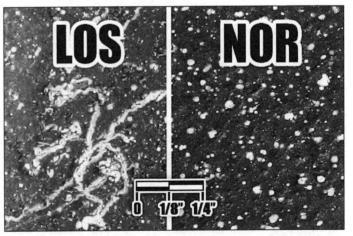

LOS- and NOR-built cars had the same base trunk paint, but the flecks of gray were different. The LOS (Van Nuys) flecks were more stringy than the NOR version. The base coat was thick and had a rough texture.

CHAPTER 4

Body shops have plenty of these work stands that hold parts for body work and painting. Some adjust to get panels at a good working height. If you're painting a lot of panels, you'll need a few extra stands.

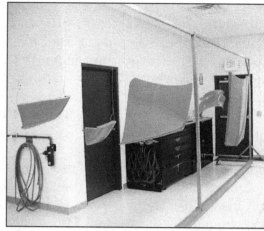

If you're painting in a high-debris environment (garage, outside, etc.), hang your parts vertically, especially when spraying basecoats and clearcoats. A panel lying horizontally is just asking for every bit of dust and debris to fall right onto it. Gravity is not your friend without a paint booth. This hanging method promotes sagging paint, so be careful! (Photo courtesy Ken Lucas/Lucas Restorations)

with more defined round speckles. The Van Nuys speckles can only be described as stringy.

This paint, especially the Van Nuys version, is extremely difficult to reproduce to factory results and nobody offers this paint in a spray can. Camaros Plus sells trunk paint close to the Norwood finish. You can also purchase many different types and colors of generic trunk paint from Dupli-Color and other sources, and the finish will be nice looking; it just won't look correct, so you need to figure out how correct you want your car.

At the factory, a human, not a machine, sprayed the trunk with a special gun after the body received its paint. Just about every panel in the trunk was covered with the spatter paint, except for the trunk weatherstrip flange, which remained body color. The underside of the package tray did not typically get sprayed with trunk paint. In fact, not much more than primer ever got onto the surface under the package shelf. That's why a lot of surface rust is usually found under there.

Spraying the trunk paint can get paint everywhere in the car because the braces between the trunk and the passenger compartment are not solid, but the braces were sprayed with the trunk paint. The factory applied the trunk paint before the car went to assembly, so none of the interior parts were in it before the trunk paint was applied. It's best to spray the trunk before the interior is installed, but if you're doing this as a project on a completed car, you should remove the back seat and cardboard panel behind the seat and above the package tray and mask off those areas before spraying. If your car is completely disassembled, we strongly suggest that you make a little effort to block the openings in the bracing to keep overspray to a minimum, otherwise the paint will go everywhere in the car.

After years of trial and error, Brian and Joe at The Super Car Workshop have been able to reproduce factory-looking results with a multiple-step process, which they have perfected and call their own "secret recipe."

Spraying Format

When you're painting, the gun mists the paint onto the surface. If you paint a partially dried panel, the new paint leaves a very rough texture. So when painting an entire car, start on one panel, go around the car, and get back to the one that is partially dried.

This is the reason why you don't start or stop painting in the center of a door, or any panel. You want to start on one end of the panel and paint your way to the end of the panel, following these steps:

Paint the right side of the roof, starting at the center, moving outward so you don't drag your gun (or anything) across wet paint. Then move to the right A-pillar.

Move to the left side of the car and spray the left side of the roof as you did the right side. Then move down the left A-pillar.

Spray the leading edges of the front fenders, lower valance, and header panel.

Start spraying in the center of the left side of the hood and move out to the top of the left front fender, making sure you paint the header panel in the same sweeping motion while painting the hood.

Spray the lower valance as you move around to the right side of the car and start spraying in the center of the hood and move out to the top of the right front fender, being sure to paint the header panel too.

Paint the right front fender, starting by spraying the radius of the fender opening, then spraying the fender at the front leading edge and moving down the fender.

Once the right fender is done, move down the side of the car to the door. Spray the right sail panel down to the top of the quarter panel, then spray the radius edge of the wheel opening and move on to the quarter panel.

Paint the top surfaces of the back of the car, and then paint the tail panel.

Spray the left sail panel, left wheel opening radius, and then left quarter. Spray the left door and move to the left front fender.

This is the basic way to shoot a car that is completely assembled. If your car doesn't have the front sheetmetal assembled, paint the body in the same manner. You don't have to worry about the front sheetmetal until the body is done. It makes sense to shoot the body and all the parts at the same time to cut down on cleanup of the gun.

Spraying Disassembled Parts

If you're going to spray all the parts of the car when they're not attached to the shell, you need some portable tubular work stands to lay the parts on while spraying them. If you have access to some tall portable racks or stands that can suspend a hood, deck lid, and doors with their large flat surfaces hanging vertically, you should use them to your advantage. Laying a hood out flat (horizontally) on a work stand leaves a huge surface just waiting for debris to fall on it. Even the cleanest paint booth has some sort of dirt or debris floating around when you least want it. With the large flat surfaces hanging vertical, the debris has a much smaller surface for gravity to pull it onto your panels. Errant debris in your paint means spending more time fixing problems.

This is less critical when you're spraying primer fillers or guide coats, because you plan on hitting those surfaces with sandpaper, which takes out most debris as it smooths the surface. Spraying basecoats and clears on vertical surfaces pays off in huge dividends.

It's a good idea to paint all the pieces in one session, so you save time on cleaning the gun, and the paint should be more uniform.

Be sure to purchase enough paint to paint the back side of the body panels. If all the panels are off the car, spend a little extra time prepping, priming, and painting the back side of the panels. This is not only good insurance against rust, but you'd be surprised at how much you can see of the back side of fenders, header panels, lower valance, and other sheetmetal after the car is reassembled. These surfaces are not as noticeable from the engine compartment when it is dirty or painted black with a spray can.

Once the car is assembled after a new paint job, ugly back sides of panels stick out like a Ford Mustang at a Super Chevy Sunday car show. You don't have to spend tons of time on them, but at least make them look decent. Whether you're going to paint them with KBS rust-preventative basecoat and top coat it with KBS BlackTop (semi-gloss) chassis paint or paint them the body color, paint the back of the panels before spending time spraying the front side of the panels. This makes sure the final and most important painted surface is facing outward on most critical surfaces.

Masking

Use large pieces of masking paper. Using a bunch of newspaper pages is a bad idea because of all the seams you need to mask—the more seams the more chances for the paper to separate while spraying air and paint on the car. Ink can easily transfer from newspaper, so no professional shop should ever use newspaper to mask a car for painting.

Use a good-quality masking tape, such as 3M Fine Line, because even doing your best job masking a car, there will inevitably be some tape that lifts a little or allows a little paint to wick under the edge.

Steps Before Spraying

Okay, now you've sanded the car and its parts for the last time, and you're ready to move all the parts into the paint booth. First, use compressed air to blow off every surface and inside every nook and cranny. Make sure the air is filtered and free of oil and water. Otherwise, you're just going to blow more contaminants onto your panels. If you're using

CHAPTER 4

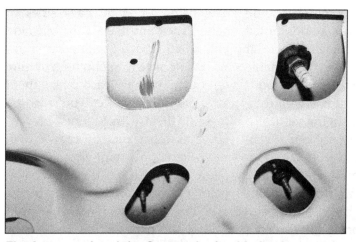

The factory painted the Camaro body with the doors and deck lid attached, but without the front sheetmetal. After the underside of the deck lid was painted, the painter had to lower the trunk lid to either inspect or paint other sections. Evidence of this is the knuckle mark in the access holes.

There has been controversy over the years about the blacked-out tail panels. Original-paint cars show that the paint finish is gloss black, not satin black.

work stands to lay parts on, make sure you spray debris off them too. Everything to go into the booth could carry debris that may end up right in the middle of your hood for everyone to see.

Move the parts into the booth for more cleaning. Once everything is in place, start by wiping all the panels with a clean microfiber cloth saturated with paint prep solvent. Follow that with a clean tack cloth, which should be loosely folded to allow it to easily do its job correctly of picking up dirt. Follow this by blowing off the surface one more time with the air. Periodically, check the tack cloth to make sure it's not clogged with debris and refold it, so it can continue doing its job properly. If you've got a lot of dust and debris in the cloth, use a new one.

Double check that all the paper and masking tape is still in place before you start spraying. By now, you realize that air shoots out of the gun and there's a good amount of walking and movement in the booth during the painting process, which can stir up dirt. One old trick was to spray the ground with water to reduce the dust. Some painters would rather sweep the floor really well than add more humidity in the paint booth, which can affect the paint's drying time.

Black-Out Paint

The factory blacked out tail panels and rocker panels for specific option codes and specially equipped Camaros. The rockers were painted black on 1967 and 1968 RPO Z22 (RS) cars. In 1969, the rockers were painted black on RPO Z21 (Style Trim Group), RPO Z22 (RS), and some 9560 and 9561 COPOs. If these cars had dark exterior colors or were Z10 or Z11 cars, the rockers were not blacked out. The SS396 Camaros received the blacked-out tail panel, with the exception of the ones with Tuxedo Black exteriors and special models, which included Z10 and Z11 Camaros.

At the factory, workers performed paint additions and final touch-ups in a place called the "paint repair shop." This department was near the end of the painting process line and did stripes, cowl paint, and black accents as well as touch-ups on insufficient paint jobs. All of these painting procedures were performed before the body component assembly. Workers on the assembly line masked the rockers, rather crudely. They didn't do a much better job with the ends of the DX1, SS, and Z28 stripes that terminated in the jambs. The tail-panel masking on the trunk jamb wasn't great, either. The factory stripes, stripe coloring of cowls, black rockers, and black tail panels were all painted with lacquer over the body paint and had a gloss finish. Most original paint Camaros have overspray and poor transition lines on stripes that terminated in the jambs.

Stripes

There are two different methods for striping your Camaro. You can

50 HOW TO RESTORE YOUR CAMARO 1967–1969

THE PAINTING PROCESS

As you can see on this 1968 RS, which has been meticulously restored to the factory overspray, the factory did not do a great job masking the jambs before the rockers on specially optioned cars were painted black. Like the tail panels, this black paint had a glossy finish.

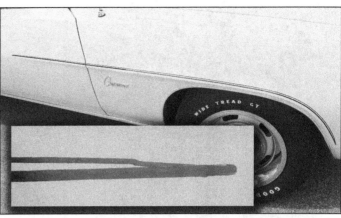

The D96 stripes accented the wheelwell openings and followed the body line that trailed off the top. The inset photo shows imperfect transition where the stripes join—they have "character" from being applied by hand on the assembly line.

either paint them on or you can use some vinyl stripes. The factory applied all the striping on the first-generation Camaros with paint. The 1968 and 1969 Camaro received the D90 stripes and thus are the exception to this standard. The fenders were painted on the 1969 models, but the door was a decal. The 1968 Camaro SS with D90 stripes was painted in front of the Camaro script and across the header panel, but decals were used behind the script and on the door as well. If you don't care about being correct you can put whatever stripe you want on your Camaro. If you have doubts about what a stripe will look like on your car, you can use 3M Fine Line masking tape to lay out some sample stripes before painting them on. A roll of tape is much cheaper than adding stripes you don't like.

You can use the Internet to search for stripe and color combinations for your year Camaro. Or you can use art programs on your computer to draw custom stripe combinations on your car or you can go old-school and draw your ideas down on paper.

If you want to go the easy route (but not always factory exact) to lay out your stripes for painting them on, you can use a stencil kit to lay out the outlines of the stripes. Stencil kits are available through stencilsandstripes.com and phoenixgraphix.com.

Factory Stripes

If you're going to use factory striping, you can refer to the measurements and illustrations in the *Camaro Assembly Manual* for the DX1, D90, D91, D96, Z10, Z11, and Z28 stripes. The stripes vary in shape by year and model, so be aware of the differences if you're trying to put the correct stripe on your car.

DX1: The 1969 RPO DX1 was more like the 1967 nose stripe except in the center of the header panel the stripes swept back in spears on the center of the hood.

D90: In 1968 the RPO D90 was introduced as a nose stripe that ran across the header panel, halfway down the nose of the fenders, and swept down the belt line of the body to about 3-1/2 inches from the rearmost doorjamb. The 1969 version of the D90 stripe was only on the nose of the front fender and turned down the side of the car on the top outside edge of the fender and the door in a spear shape. The 1969 version is more commonly known as the "hockey stick" stripe.

D91: The RPO D91 was a 1967 and 1968 stripe. It's the common nose stripe on the header panel and front fenders, which is referred to as the "bumble bee" stripe.

D96: The 1968 D96 was pinstriping which was painted on the beltline peak from the front fender to the rear quarter panel. In 1969 the D96 pinstriping was only on the accent-peak body lines, starting at the front of the wheel openings and following the accent-peak rearward.

L48: In 1968 the RPO L48 modified the nose stripe to include a pointed stripe that ran from the nose down the belt line toward the rear wheel opening.

Z10, Z11, and Z28: All three first-generation Camaro Z28s and the 1969 Z10 and Z11 RPOs had the wide stripes surrounded by a smaller

CHAPTER 4

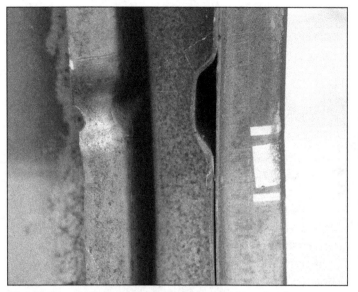

The factory painted the body separate from its front sheetmetal, which was also not assembled. The stripes reach all the way to the gaps between the panels. This has been documented by disassembling survivor cars with factory paint. (Photo courtesy of Frank Arone)

The Z28, Z10, and Z11 stripes stop on the trailing edge of the spoiler. The factory painted the stripes after the rear spoiler was installed (inset, right). The second set of emblem holes were covered/sealed with black gaffer's tape (inset, left). In this rare case, the tape was painted over and visible with the spoiler installed. (Photo courtesy of Frank Arone)

1/4-inch stripe that ran from front to rear on the top sheetmetal, excluding the roof. If a Z28 had a rear spoiler, the stripes carried over the top of it and stopped at the bottom of the curve on the back side. All the Z10s and Z11s had the D80 front and rear spoilers (both were part of the package) as standard equipment, and the stripe was the same format as on the Z28.

Z21: The small painted pinstripes on the 1967 Camaro were designated as RPO Z21. These stripes ran from the front to the rear of the body on the side below the upper ridge of the body line.

Painted Stripes

For painting factory stripes, you can buy stencil kits to properly lay all the stripes on your car. The benefit of using paint is that you can paint the stripes the correct color with a glossy finish. If you don't care about the three base factory stripe colors of black, white, and red, you can personalize your car with any color or finish.

For an accurate factory restoration, you need to paint the stripes on the body with a single-stage paint just as the factory did. The factory painted the front sheetmetal separate from the car and in pieces. That means when it painted the SS nose stripe, the header panel was separate from the fenders, and if it was painting Z28 stripes, the header panel, hood, and cowl cover panel were not attached to the car. (Proof of the header panel being painted separately has been documented on original-paint cars.)

With the header panel removed, the stripes were painted in the jambs where the panel was mated to the fenders. The deck lid was attached to the body when the shell was painted at the factory, so the factory-applied Z28 stripes were painted with the deck lid attached.

Factory Z10, Z11, and Z28 stripes were not painted under the spoiler. If your Z28 was factory equipped with a D80 rear spoiler and you didn't order "stripe delete," the stripes went over the spoiler but ended at the bottom of the curve on the back of the spoiler. If the Z28 stripes were applied at the factory but the car did not have a D80 spoiler, the stripes were painted all the way down the deck lid to the trailing edge and stopped at the deck lid jamb. If you have a Camaro with a D80 spoiler with the Z28 stripe and paint on the trailing edge of the deck lid, a dealer or a customer added the spoiler and applied stripes to the spoiler. All Z10- and Z11-equipped Camaros had the orange Z28-style stripes.

Laying out your own stripes by hand can be really tough because it's a very tedious job making sure the edges are straight and the stripes on the left and right side of the car are symmetrical. Take accurate measurements and make sure everything looks spot-on before you spray color on the car.

THE PAINTING PROCESS

This Yenko was restored once before. Fortunately the previous restorer did not prep the hood well and a faint outline of the original decal could be seen. The restoration decals (white) were not the same shape as the originals (outlined with tape). (Tony Lucas Collection)

Rust under the vinyl top is expected after 40-plus years of exposure to regular and even non-regular service. The rust associated with the vinyl top is typically severe around the rear window channel.

The professional painters we talked to won't use anything but 3M Fine Line masking tape to lay out their stripes. A clean surface is key to professionally laying the stripes. Confirm the tape is completely adhered to the body, especially in the corners where masking tape typically overlaps. This cuts down on the paint making its way under the tape and leaving you with an ugly mess to fix. There's also a specific method to removing the masking tape.

After painting the stripe color, gently press on the paint that is laid over the masking tape with a clean finger. When the paint doesn't show a fingerprint, the paint is dry enough to pull the tape off the car at a 90-degree angle from the panel. If you let the paint dry too much you can get a ridge of paint built up that can damage the stripe when removing the masking tape.

If you don't care about being factory correct, with today's clear top coat paints, you can easily paint stripes on a car before the clear goes over the basecoat with very few raised edges. With some extra attention and extra coats of clear, you can completely eliminate the raised stripes' paint edges. In the past, before clear topcoat paints were readily available, it was almost impossible to paint stripes without a thick edge on them.

Vinyl Stripes

If you're considering installing vinyl graphics on your car, you can use kits like the ones from stencilsandstripes.com and phoenixgraphix.com to get the factory shape, or you can get custom stripes made. The technology of vinyl graphics has come a long way in the past few decades. High-quality modern materials last much longer in the elements than earlier vinyl graphics. Vinyl stripes come in white or flat or matte black.

The benefit of using vinyl is that you are not necessarily locked into a long-term commitment with them if you don't like them. You can't say the same about paint. Keep in mind that the sun does bleach paint. If you remove vinyl stripes after even a few months in the sun, you run the risk of having noticeably darker ghosted stripes left in the paint. If you wait years, you'll have some very obvious stripes left in the paint. With the computer technology used by vinyl graphics companies, you can get just about any shape of stripe cutout, and they will match from side to side. This is especially helpful when creating intricate designs.

Follow the installation instructions that came with your stripe kit.

CHAPTER 4

Here are a few good tips that companies sometimes leave out of installation instructions:

- With all your measurements determine exactly where you want the stripes to go.
- Take a few pieces of masking tape and stick them to the body 1/4 inch wider than your stripes. This give you a guideline to make sure the vinyl ends up centered between the masking guides. Without these, you're flying blind and you're bound to get the decal in the wrong spot.
- Measure two or three times before taking the backing off the decal. Always apply adhesive graphics on a cool surface; never work with hot surfaces for this kind of task.
- Never put a decal or vinyl on a dry surface, there should be instructions accompanying your kit on what water and soap solution (typically 98-percent water and 2-percent dish soap) to spray on the surface before laying the accent on the panel. If your panel is hot, the soapy water solution evaporates before you're able to apply the graphics and you have a huge mess on your hands. The water and soap solution allows you to move the graphic into proper position before it sticks to the panel and also, so you can use a squeegee to gently force the air bubbles out from under the graphic.
- Start to squeegee the air bubbles and water out from under the center of the decal and work your way out to the edges. Starting at the edges locks the edge down to the panel and you won't be able to push air bubbles out of either side. Always try to get the air bubbles to the closest edge. In other words, don't try to push an air bubble 2 feet across a hood if you can push it 2 inches to the outside edge of a decal.

Vinyl Top

The other exterior option that's installed on the body after a car is painted is the vinyl top. Vinyl tops are thought of as bad because water gets trapped under the vinyl and inevitably causes rust to form around the window channel and trunk filler panel. Many people do not like the potential rust problems and poor life expectancy associated with a vinyl top.

In humid climates, the rust damage associated with the vinyl top is typically moderate to extreme and expensive to repair. Even dry climates can wreak havoc on the panels under old vinyl tops. The seam between the filler panel and the quarters must be sealed properly on the top surface and in the trunk gutter in order to keep rust from reoccurring in the future. With the exception of cars being restored back to original, almost all cars get their vinyl tops removed completely during a restoration.

Powder Coating

If you don't care about factory finishes and are looking for a durable finish for your parts that's a lot tougher than a coat of spray paint, you should look at powder coating. Some can mimic the factory finish, but the coated thickness can wash out some textures on cast parts. There is a huge spectrum of powder coat colors and finishes available. For years, if you wanted to have an extremely tough and durable powder coating applied to your parts, you had to spend a bunch of money having a specialty shop do the work for you. Eastwood provides many automotive restoration and customization solutions.

One of those solutions is a simple and effective powder coating system you can use in the convenience of your own shop or garage. The device is the HotCoat powder-coating system. You need the HotCoat kit, powder colors, an old oven you don't prepare food in anymore, a compressor that can deliver 5 to 10 psi, 110-volt electricity, and a little space to coat your parts. Eastwood also sells special tape so you can mask the surfaces you don't want to coat, and special plugs to keep the powder out of threads and little holes.

The simple-to-use Eastwood Hot-Coat system makes it easier to coat parts than to paint them. Unlike with paint that you have to worry about applying too much and having a sag or a run, you can't really apply too much powder coating. Simply follow the process of energizing the part on a wire rack, blow the powder onto the part, unhook the power to the rack, carefully move the rack to the oven, cook the part, allow the part to cool, and install it. Now you have a part that isn't easily scratched or chipped, so it will look great for years to come. The beauty of having a durable finish also allows the surface to easily be cleaned so your project is much more fun to work on and show off to your buddies. The powder overspray on the floor is also very easy to clean up with a broom and dispose of.

You can still have a shop coat your parts, especially since some parts are too large to fit in your oven. Most powder-coating shops also

THE PAINTING PROCESS

have a much larger selection of colors than Eastwood's 92 (at the time of this writing). Professional shops also typically have more than the four colors Eastwood currently offers in high-temp coatings.

Scuff and Buff

After all the bodywork, primering, and painting has been finished, there's one crucial step left: scuffing and buffing the paint (a.k.a. color sanding and polishing). No matter how good a painter is, the paint finish will have a bumpy and uneven condition similar to the peel of an orange, which is commonly referred to as "orange peel." As long as enough clearcoat has been sprayed on the paint, color sanding and buffing can turn an extremely orange-peel-laden paint job into an amazing paint job. Basically, the color sanding process uses ultra-fine sandpaper to smooth out all the tiny ridges and peaks of the clearcoat and then you use rubbing compound to remove the sanding marks. After that you are left with ugly swirl marks in the paint that must be polished out, leaving the surface very flat, smooth, and beautiful.

Remember, there must be enough clear over the basecoat. If the clearcoat is too thin or too much is removed, the following process will destroy your paint and car will have to be repainted. The process here is what worked best for the car shown. The same shop that painted the car also color sanded, buffed, and polished it.

Paint is very expensive and many shops lack integrity and pride in their work. Many are looking to save as much money as possible, and apply a thin clearcoat when a much thicker coat needed to be applied. The benefit of having the same shop perform all the tasks from start to finish is that they will be sure to lay enough clear over the basecoat, knowing that its next step is to sand and process the paint. The shop's scuff-and-buff technician should also know how much clearcoat he has to work with because his shop painted your car.

If you take a new paint job home and sand or buff through the clearcoat, the shop is fully released of liability and you have to get the panel completely repainted on your own dime. You can't just spray more clearcoat over the area. The basecoat is very thin, so when you sand through the clearcoat, it's more than likely you will sand through the basecoat into the primer. Another important tip is not to color sand, buff, polish, or apply any chemicals on a clearcoat for at least 120 days if it was not baked on with the proper equipment at the paint shop.

A moderate-quality paint finish requires about 10 hours of polishing and buffing from start to finish, when performed by a trained professional. For show winning results, a seasoned professional usually spends 40 to 60 plus hours scuffing and buffing an entire car. If you're doing it yourself for the first time, expect to double or triple those numbers. Also expect to spend about $200 on polishing and buffing pads, wet or dry sandpaper in multiple grits, sanding blocks, tape, compounds, polishes, detail spray, and wax.

Don't forget you'll need to own or borrow a 7- or 9-inch variable-speed right-angle polisher/grinder. Make sure you get a variable-speed polisher, because they are much easier to control. A variable-speed polisher is less likely to burn through (completely polish through) the clearcoat, which requires repainting the panel. It's very easy for an inexperienced polisher to burn through the paint during this process.

I'm not trying to scare you from doing the job yourself; I'm trying to explain the challenges. Therefore, you'll know what to expect and you won't be so surprised that professionals charge so much money to scuff and buff. You're paying for their experience, hours, materials, and tools.

The process of color sanding, buffing, and polishing makes a huge mess. Expect to use a bit of water while color sanding—it's best to do this on concrete, not dirt. The variable-speed polisher is going to fling rubbing compound and polish all over the place and cleanup of this stuff is hard once you're done, especially in doorjambs and under the hood. Covering the tires or even completely masking the fenderwell openings greatly reduces the post-polishing mess. Most professionals tape the gaps between the panels and the window trim and weatherstrips. If your car is assembled, it's easy to damage the body and window trim, emblems, door handles, miscellaneous trim, and window felts (a.k.a. cat whiskers) with the sandpaper and spinning buffing pads, so taping them is important.

You should soak your wet or dry sandpaper in a bucket of water mixed with a little Ivory liquid dish soap. After many years in the business, professional Jon Lindstrom of Best of Show Coach Works, in San Marcos, California, has found Ivory to be the easiest on his hands. He also works with waterproof bandages on his thumb and pinkie finger because, as you'll find out, after a few hours of color sanding your car you no longer have fingerprints. Not long after that, you'll completely sand through your skin.

CHAPTER 4

Finish Sanding and Buffing the Paint

Rubber sanding pads or squeegees come in different firmnesses. The firm ones knock down stripe edges and orange peel faster and more evenly, but the softer ones follow body contours better. Keep the paper and working area wet by constantly feeding soapy water from a sponge and keep rewetting the paper in the bucket to keep it saturated. (Photo courtesy Steven Rupp)

In order to perform the whole task, you need wet-or-dry sandpaper in different grits, sanding pads, buffing and polishing pads, compounds, polishes, detail spray, wax, tape, Ivory dish soap, a sponge, a bucket, and a variable-speed polisher. All products and top-quality sandpaper shown here are Meguiar's, which makes it nice for one-stop shopping. (Photo courtesy Steven Rupp)

Sanding blocks come in different shapes for contours. Jon has a large collection of different blocks, and this one worked great on the hood scoop. When he doesn't have a shape that works or in tight spaces, he resorts to small flexible sanding pads while being careful not to sand through hard edges and peaks. (Photo courtesy Steven Rupp)

1 Soak your wet-or-dry paper in a bucket of water with a little Ivory soap mixed in. Best of Show Coach Works employee Jon Lindstrom has begun sanding with 600–grit paper wrapped around a rubber sanding pad. If you have five thick coats of clear over the stripes, you may have enough paint to sand the clear down to where the ridges on the borders are gone. In this case you may want to start with more aggressive 400-grit on the stripe edge and quickly move to 600-grit. (Photo courtesy Steven Rupp)

2 Notice that the partially sanded area shows where the high spots of the orange peel are starting to be knocked down. The rest of the sanded area has been smoothed out. (Photo courtesy Steven Rupp)

THE PAINTING PROCESS

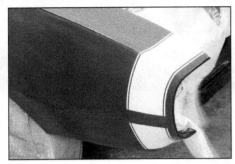

3 Once the fender has been sanded, it's time to switch to 1200-grit paper with a softer sanding pad. Notice the ridge on the top of the fender is still dark; that's because Jon hasn't sanded that part. In fact, most professionals run some striping tape along the peaks of the body to ensure they don't sand them too much by accident, but Jon has the skill not to sand them. We suggest you use the tape. (Photo courtesy Steven Rupp)

4 Next, Jon lightly sands the peak by hand (without a sanding pad) with 1000-grit and a lot of soapy water. Be very delicate and don't spend a lot of time here since the peaks rarely have as much clear as the rest of the panel because of the pull of gravity during the paint-drying process. Sand through the clear and you'll have to repaint the panel. (Photo courtesy Steven Rupp)

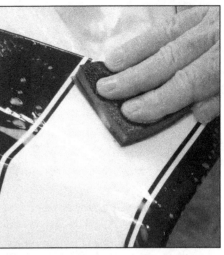

5 Sand the fender with 2000-grit paper backed with a soft rubber pad, still keeping the soapy water as a lubricant. Don't sand the peak more than to just remove sand-scratches. Then switch to 3000-grit paper without a backing pad. (Photo courtesy Steven Rupp)

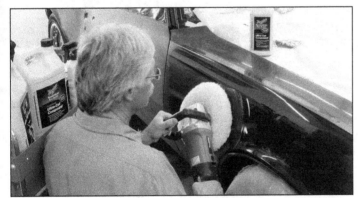

6 You can overload the pad with compound and burn through the paint quickly, especially if you're using a high-speed polisher. First use a pure-wool 8-inch cutting pad with a couple lines of Meguiar's Ultra-Cut compound with the lower speed on your variable-speed polisher. Friction and speed heat the paint and burns though it. With the scratches gone there are now finer prism-effect swirls in the paint. Switch to the 8-inch foam pad and Swirl-Free polish. With a microfiber towel wipe the panel with Final-Inspection spray by hand to clean the panel so you can check for imperfections that still need attention. (Photo courtesy Steven Rupp)

7 Shown here are the orange peel before (left) and the finished product (right). Once the whole car looks great, apply a good coat of Meguiar's wax to protect your time and money investment. (Photo courtesy Steven Rupp)

CHAPTER 5

ENGINE REBUILDING

All 1967–1969 Camaros came equipped with an L6 (6-cylinder) or a small-block or big-block V-8 engine. Most owners rebuild a small-block or a big-block, and a few enthusiasts rebuild a 6-cylinder. Therefore, in this chapter we focus on the small-block and big-block engine restoration and rebuilding process.

This book covers the assembly process after receiving the machined parts from the machine shop, visually inspecting them, and double checking clearances.

It does not cover the process of choosing the correct parts for your application, inspecting the old parts to find out what may have necessitated the rebuild, machining the parts, balancing, block cleaning and prep, or some of the finer details of engine assembly, starting the engine, and troubleshooting.

For complete instructions on rebuilding small-block Chevy engines, pick up a copy of the CarTech title *How To Rebuild the Small-Block Chevrolet* by Larry Schreib and Larry Atherton; for the big-block Chevrolet engine, *How To Rebuild the Big-Block Chevrolet* by Tony E. Huntimer. These books cover all the finer details of rebuilding engines from inspection, precision measurements, parts selection, machining, block prep, pre-assembly, and final assembly and have complete chapters covering each aspect with step-by-step photos.

Unlike some other automotive manufacturers, Chevrolet engineers have helped us by designing the small-block and big-block very similarly, in the respect of engine and transmission mount configurations, overall engine assembly design, flywheels (or flex plates), starters, oil filters, firing order, crankshaft configuration, and distributor design. But there are differences, which include head port design, valvetrain layout, and of course, physical size.

Because the 454 engine didn't come out until 1970, the difference between internal- and external-balanced rotating assemblies won't be addressed. All references to first-generation Camaro engine crankshafts and harmonic dampers are internal-balance only. (For external-balance components and most performance parts consult the previously mentioned CarTech books.)

So many Camaro owners burn a lot of calories on the outside of the car, but neglect to spend a few bucks on paint to make the engine look good. Even if you modify the internal components, the outside can look stock. This 427 was restored back to original finishes by SCW. (Photo courtesy Tony Lucas collection)

ENGINE REBUILDING

Selecting a Machine Shop

You need to find a professional machine shop that's experienced in rebuilding Chevy V-8 engines. In essence, do your research, get a good recommendation, and develop a good relationship with the shop. Many horror stories have surfaced about parts being lost, misplaced, or ending up in the hands of other customers, and you want to avoid this. Every machine shop has made a few mistakes, costly or minor, simply because the employees are human after all. At every machine shop I've been to, there are at least 20 or more disassembled engines in the building at one time.

To help keep the machine shop employees from mixing up your parts with another job, you can use a scribe, stamp, or etching tool to add a personal and distinctive mark on your large, important parts (heads, block, intake, and crank) and on non-critical gasket or machined surfaces. Make sure they are surfaces that won't be visible once the engine is assembled. Point these marks out to the machinist when you deliver the parts, so he or she can clearly identify your parts.

Note the VIN and assembly stamps on the front pad, casting numbers, casting date, and other numbers and letters stamped or cast into the parts before you take them to the machine shop. You can take pictures of them, but having them written in a book is a little more secure, unless you keep good backups on your computer or print out all your pictures. The most important numbers you should record are the VIN and assembly stamps on the front ID pad in front of cylinder number-two because as soon as the machine shop "decks" the block, those stamps disappear or at least become almost unreadable, depending on how deep the factory stamps were made. Some machine shops deck the block without asking or thinking twice about it. In one sweep of the machine all the data that can prove that your engine belongs to your car, the build week, the engine plant, and what kind of transmission was in your car will be gone forever.

When you drop off your parts at the machine shop, clearly communicate the machine work you want done, so your parts are machined to your requirements and mistakes are avoided. Also go over the things you don't want done to the engine. Just because you don't tell them to perform a task doesn't mean that they won't take it upon themselves to do it on their own.

Identification

The engine came with sort of trim tag for the drivetrain. All the info is on the ID pad in front of the number-two cylinder. Unfortunately, this detailed information is located on the deck surface of the engine, which is commonly removed if the head surface was machined to fix some defect during a rebuild. If all the information is still on the pad, you can find out which engine assembly plant built it, what month and day it was built, the engine size and HP rating, what transmission was originally behind it, and whether it had factory smog equipment.

A partial VIN is also on the ID pad of the vehicle, so you can tell if it's the original engine for the vehicle, but only in some circumstances. In 1967, the partial VIN (last eight digits: i.e., 7L100001) was typically only stamped on the pad on performance-optioned Camaros. Federal law starting for the 1968 model year required cars to have the partial VIN plus the first digit of the VIN number, which was a "1," to precede the last eight digits (i.e., 18L301507) stamped on the drivetrain (engine and transmission). During the 1969 model year the partial VIN can be found either on the front ID pad or on the rough-cast engine-to-transmission boss over the top of the oil filter.

The speculation behind the location change is because the alternator location changed from the driver's side of the engine to the passenger side, where it covered the ID pad. If the ID pad has been machined or the numbers have been removed for some reason, you won't have any application info specific to your car. The only info you have is the original casting numbers elsewhere on the engine parts, which proves nothing more than casting dates and really doesn't add to the value of the car like the info on the ID pad does.

Casting Numbers

Most of the larger cast parts on the Camaro have an identifying seven-digit number cast into them, which is the part number. Other numbers and letters cast into factory: intake and exhaust manifolds, heads, blocks, and crankshafts are casting date numbers, lot numbers, nickel content percentage, mold numbers, R and L (right and left in the case of exhaust manifolds), ignition firing order, application designation, and other unknown casting plant identification.

If the numbers have been ground off the ID pad, you can at least check the casting numbers to learn more general information about what year, size, and horsepower rating the block may have been used for.

Record the numbers stamped on the pad in front of cylinder number-2 with a photo and in your notebook because a machine shop may machine it off. Important information is on this ID pad. It contains the partial VIN (left) that matches the VIN of the car it came from and the digits VI2I5MS. The letter "I" was used as the number 1 in engine codes. The letter "V" meant it was a Flint-built engine. Code I2 meant assembly of this engine was done in December and I5 meant on the 15th day of the month. MS meant this engine is a 295-hp 350 with a manual transmission. If the transmission codes match this car and engine, the car would be considered "numbers matching."

In mid 1969 the factory started stamping the block above the oil filter on the transmission mounting boss with the car's VIN number. For 1968, federal law required the use of the first digit (1) followed by the last eight digits (i.e., 9N607195).

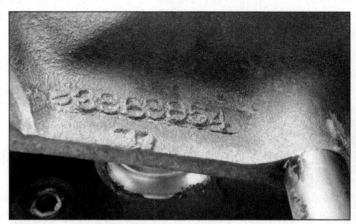

The block casting number is located on the transmission flange behind the number-7 cylinder. This one is 3969854, which was a late 1969 396/325 or 350 hp, if it's a two-bolt main; or it was a 396/375 hp, if it's a 4-bolt main. This block was also used as the 402 block from 1970 to 1972. A quick check of the casting date revealed March 19, 1971, so this block is a 402 (a factory .030-inch over 396).

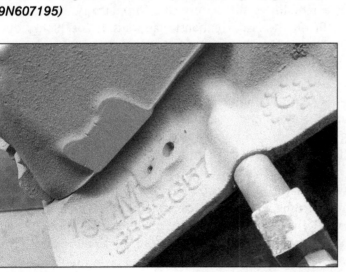

If you're interested in using the correct heads on your small- or big-block, check the casting dates and numbers that are located under the valve covers. There are some external distinguishing marks, like the 1967 and 1968 1.94 and 2.02 (refers to the intake valve diameter) heads with the double-hump casting mark (a.k.a. camel hump). These 1967 and 1968 small-block heads didn't have accessory mounting bosses in the heads. The 1969 small-block heads required the bolt holes in the heads because that was the year Chevy moved the alternator to the passenger side and required the holes for mounting. The 1969 1.94 and 2.02 heads had different casting shapes, one of which was a short double hump.

ENGINE REBUILDING

Some blocks were used for multiple years and even different engine sizes. For instance, an engine block with the casting number of 3969854 was used for 1969 396 Camaro applications, and it was also used for multiple car platforms that used the same block for 1970 through 1972 402 applications.

We confirmed the 396 block was actually a 402 because the casting date was C1971. The coding convention for the block, heads, and intake manifold is the same format for the three years we're covering in this book.

The "letter" stands for the month, where C is the third letter in the alphabet, so it would be the third month of the year, which is March. The second and third digits are 19, which is the day of the month, and the fourth and fifth digits are the year, so the engine was cast March 19, 1971. In 1970, the factory stopped building 396s and started boring them .030 over to 402 ci, so this was definitely a factory 402 block that we were decoding.

On a side note, the 1970 and 1971 Camaros still had 396 badges on the fenders, but the engine was a 402. The 396 label had a lot of marketing value, so Chevrolet kept the 396 for a while.

Inspecting and Preparing Parts

Physically inspect every part before you turn a wrench. Check for burrs, cracks, and gouges in every surface from the tips of the valves to the bearing saddles in the block. Clean out every thread in the block and crankshaft to make sure the threads are clean and in good shape. Stick your finger into the lifter bores and make sure there aren't any burrs

This 1968 engine came out with a trim-tag build date of 12E, which means December in the fifth week of 1967 (the model year starts with August or September of the previous calendar year). The casting date on the block is L137. The L is the 12th letter of the alphabet, so it was cast in December on the thirteenth day of 1967. There can typically be a couple-week gap in casting dates between engine blocks, heads, and intake manifolds. The intake was cast L17 (December 1, 1967). The intake and block casting dates closely match and they both match closely enough to the trim-tag build date.

Never wedge any tool between the head and block surface. Remove all but two head bolts, which should only be threaded in a few turns, before trying to break the head loose. This prevents the head from falling on your foot. A wooden hammer handle in an intake port can be held when prying the heads loose. If the lifters are mushroomed and won't pull out of their bores, don't force them because you'll damage the bore. Simply pull the lifter up and put a zip-tie around it to keep it elevated, so you can pull the camshaft out. Drop the lifters into the engine, or you can cut a plastic pipe in half, stick it in the cam bore, and drop the lifters into the pipe. Valve locks can burr the lock groove on the valve so you can't pull it out of the head. If this happens, file down the burr and the valve will slide out without damaging the guide. Check for a ridge near the top of the cylinder. Use a reamer tool to remove the ridge and the pistons. However, if the ridge is too deep, you probably won't be able to get the pistons out of the bores, and trying to remove the pistons could result in block damage. If this is the situation, the machine shop should remove them.

CHAPTER 5

Taking precise measurements is critical for building a strong-running engine to last for a long time. Unless you actually watched the machinist perform all the key measurements, you'll need to check all the measurements as insurance. All responsibility of the engine build is on you. If the machine shop made a mistake, it's up to you to find it before you turn a wrench. These specialty tools came from Goodson and are required: dial bore gauge, dial indicator (with magnetic stand), caliper, outside micrometers, and straight edge. Some tools can be used for more than engine building, so you may put them to use on other projects. Building a performance engine requires additional precision tools, such as a cam degree kit, valve spring height checker, vise-mounted spring compressor (at minimum), and magnetic deck stand with dial indicator.

Using Plastigage is the back-up method for measuring the crankshaft main bearing and rod bearing clearances. It's a precision strip of a special plastic that you crush on a bearing surface to measure the bearing clearance. To get a correct reading, the bearings and the surface they install into must be completely clean and oil free. Once installed in the rod or block, leave them oil free because any debris or oil causes an incorrect measurement. The crankshaft needs to be oil free also. During this process, don't turn the crank at all because it damages the Plastigage and you have to start over, and you can damage the bearing and the bearing surfaces. If you're checking the main bearings, install all the bearings in the block. Carefully set the clean and oil-free crank onto the bearings. Put a little strip of Plastigage across the bearing surfaces. Torque the main caps in the block with the bearings in them. Remove the main caps and check the width of the squished Plastigage on the Plastigage sleeve. Checking the rod-bearing clearance is tougher because the crank tries to turn while doing so. You can check the rod bearing clearance further along in the engine build. It's much easier to check after the crank is installed, the main bearings are lubricated, and two pistons (without rings) and assemblies are torqued in place on one rod journal at a time with the rod bearings dry. Installing both on the journal keeps them from twisting too much, and you can check side clearance at the same time. To keep the crank from turning, stand in front of or behind of the engine (not on the sides) and torque and loosen in line with the length of the crankshaft.

Only use paper shop towels to clean or wipe your engine components. Professionals never use any kind of cloth towels during engine assembly. Threads and cloth fibers get into the engine and cause bearing failure later because they don't break down in the oil. Paper towel fibers break down in the oil. Our choice for paper towels is the blue Scott Shop Towels. They are very durable and absorb engine oils better than your typical paper towel. In order to thoroughly clean a gasket surface, wipe the surface multiple times. If you don't clean the oils out of the porous cast-iron gasket surface, you'll be getting an oil leak very soon.

ENGINE REBUILDING

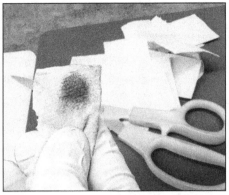

The best way to get the surface clean is to use non-chlorinated brake cleaner. You're going to use a lot of towels, unless you follow our lead on this. Spend a few minutes cutting your blue shop towels into 20 or 30 little squares about 2x2 inches. Do a basic clean of the surface with a full-size paper towel and carb cleaner. Then grab one of the little squares and squirt a little carb cleaner directly on the paper towel. Wipe the gasket surface before the carb cleaner dries and I guarantee the towel will have oil on it. Do this to the entire gasket-sealing surface. That also goes for all gasket surfaces, from the oil-pan gasket surface to the timing-cover gasket surface to the intake-manifold gasket surfaces. All of them need to be as free of embedded oils as possible, or they'll rise to the top and cause a leak. (When I refer to cleaning a surface, this is the method I'm referring to).

Working on your Camaro, especially the engine, involves the use of chemicals that can cause serious illnesses and cancer. Protect your skin from these unhealthy chemicals. Use nitrile work gloves from the auto parts or home center store. Don't use silicone work gloves; they don't stand up to the harsh chemicals in solvents, fuel, and brake cleaners. The harsh chemicals still attack the nitrile gloves, but not nearly as fast as silicone gloves. With the correct size, you should not lose dexterity. If you're wearing gloves when installing bearings, make sure you don't get the glove caught under the edge of the bearing. If your glove gets snagged under the bearing, there will be a visible hole in the glove. If this happens, pull the bearing again and reinstall it without the nitrile chunk under it.

Our 1968 small-block project looks close to a factory-original engine once it has been assembled. It has cast-iron exhaust manifolds as well as a correct cast-iron intake manifold. We decided to use high-performance aftermarket parts inside the engine, such as a Speed Pro: roller hydraulic retrofit cam and lifters, matching valve springs, guide plates, hardened pushrods, stainless-steel valves, stainless-bodied roller rocker arms, double-roller timing set, Mahle hypereutectic pistons, Victor Reinz gaskets, Clevite coated bearings, Perfect Circle rings. From Milodon we used: lifter-valley screen kit, stock-appearing oil pan with built-in windage tray, one-piece oil pan gasket, high-volume oil pump, pump shaft, pick-up, lifter-valley screen kit, and ARP bolts. We're reusing the original forged crank and reconditioned connecting rods.

CHAPTER 5

or rough surfaces in them. The lifters have to be able to rotate freely in their bores as well as move up and down. If they don't rotate freely, they will stay in one radial position and grind into the cam and cause the cam lobe to go flat and the metal particles to travel throughout the engine and its bearings.

Don't use a bolt tap to check the threads. The tap is meant to cut new threads from damaged ones. When it cuts new threads, it removes material from the thread, which decreases its strength. The next time you try to torque a bolt in that hole you may pull out what's left of the thread and have to install a thread-repair insert. There are specific "thread chaser" taps that only clean the threads instead of cutting new ones. ARP Fasteners sells these special chaser taps. The machine shop can clean and check the threads for you for a fee if you don't feel like purchasing the equipment or spending the time.

Engine-Rebuilding Lubricants

Since 2007, oil companies have drastically reduced the zinc and phosphorus content in the typical over-the-counter engine oil, so the oil meets environmental regulations according to industry standards. Without these and other additives, flat-tappet camshafts in older engines are going flat in record numbers. The additives keep the high-load friction surfaces of the cam lobes and the lifters from grinding each other into oblivion. The majority of the cars on the road these days have roller cams or their valvetrain design is much less friction-based, so the missing additives aren't causing problems with them.

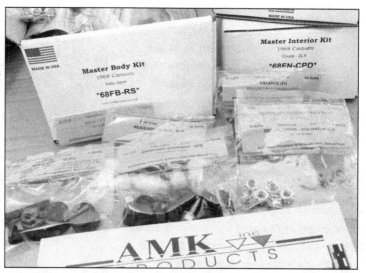

If you go with a strictly stock engine restoration all the way down to the bolts, you can get reproductions from the typical parts sources, which are partially correct. However, if you want the exact bolts, shop with AMK Products for small- and big-blocks. It offers all the different bolts for aluminum and cast-iron intake manifolds and other specialty engine hardware that is very specific for each application. The heads have the correct shoulders, head stampings, and bolt finishes. The 1965 through 1974 big-block factory head bolts have a little "M" stamped in one segment of the X stamp where some reproduction and all aftermarket bolts do not have the correct markings. It all depends on how correct you want to be.

Verify all the machine work before assembly. Take all of the proper measurements to confirm that the machine shop didn't make any mistakes. As soon as you turn one nut or bolt you've automatically taken responsibility for everything. If the cam goes flat or you spin a bearing on fire-up, it's all on you. If you're not comfortable with this responsibility, pay a couple hundred more dollars and have the machine shop assemble the whole thing from intake to oil pan.

ENGINE REBUILDING

These old heads required lead to keep the cast-iron valve seats lubricated. The lead in gasoline has been gone for many years now, and the valves start hammering the soft valve seats into oblivion. In the picture you can see the valve boring its way through the seat. The only way to combat this problem is to machine into the original seat area and install a new hardened seat that accommodates today's fuels. This head is pretty much junk because the seats are so far gone that the shop would have to drill into the water jacket to replace the extra-deep holes.

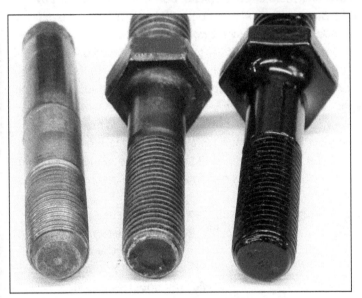

Unlike the big-block heads, the small-block Chevy heads have pressed-in rocker studs. If you are upgrading to anything higher than 350 hp, consider having the machine shop install some ARP screw-in rocker studs and guide plates. The pressed-in studs tend to pull out of the head from severe stud deflection caused by increased cam lift and valve spring pressure. The old studs need to be pulled out and the head machined and threaded to accept the screw-in studs. If you add roller rocker arms that require the posi-locking nuts on a big-block Chevy engine, install new screw-in rocker studs that have a flat tip on the studs because the locking Allen screw in the center of the rocker nut needs to tighten against a flat surface so it doesn't come loose. When you upgrade to guide plates, also use hardened Chromoly pushrods or the guide plates wear through them, causing a lot of metal debris in your engine.

We decided to keep the original castings and replace the valve seats with hardened seats. In the process, we decided to increase the exhaust valve and seat size from 1.50 to 1.60 inches to help get more exhaust gasses out of the cylinders. Here, Bob Gromm of Gromm Racing Heads, Balancing & Engine Machine works his magic on our heads. The valves we chose are Sealed Power forged-stainless-steel valves to get away from the original cast valves. The valves are one-piece design, swirl polished, with a hard chrome stem and hardened tip.

CHAPTER 5

Engine Assembly

1 If your pistons require pressed-in wrist pins, the machine shop can install them easily with the correct equipment. The end of the connecting rod is heated to a specific temperature so the rod strength is not compromised. Then it's set into the center of the piston and the wrist pin is pushed in a specific amount. Trying this at home can go wrong in a hurry. If it's done incorrectly, you ruin the rod, wrist pin bore, and wrist pin. There is a wrong and a right piston-to-rod orientation. This procedure is usually done after all the machine work is complete. If you have full-floating performance pistons, assemble them at home because there's no pressing necessary for interference fit parts. The wrist pin is not locked to the rod, and wire spring clips or spiral locks hold the pin in place.

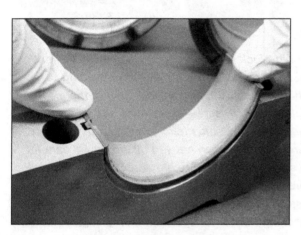

2 Before installing engine bearings, check the bearing saddles of the block and main caps for nicks, burrs, and gouges. If any burrs keep the bearing from sitting completely down into the block, you may be able to knock it down by dragging the blade of a knife across it. Don't gouge or dig at a burr. If it's too large or becomes a questionable task for you, contact your machine shop. The saddle should be completely free of oil and debris. Even if it looks clean, wipe the saddle with a little piece of paper towel with carb cleaner on it. Press the bearing into the saddle evenly. Make sure the tang on the bearing lines up with the tang relief in the saddle. If it doesn't line up, remove the bearing and start again. Don't shove the bearing sideways or tweak it, or you'll damage the bearing and cause engine failure later.

 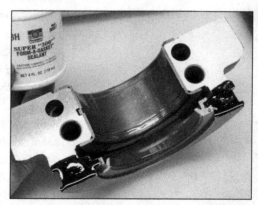

3 Clean the bearings, bearing saddles, and registers with the carb cleaner and shop towels. Install all the grooved main bearings in the block. Install the bearings in the main caps too. Put a light film of oil on the cam bearings (while you can still reach them), on the main-bearing surface only (not all over the rest of the parts), on the main bearing journals, and on the rear main seal area of the crank. Install the rear main seal completely dry, except for a light film of oil on the lip that rides on the crank. Add a light coat of Permatex Super 300 on the main cap surface (shown). As for the rear main seal position, we prefer clocking the seal about 1/4-inch off the parting line of the main cap. Carefully lower the crank into the block, making sure not to damage the rear main bearing with the thrust flange. Make sure the main caps are in the correct order and direction, and seated in their registers. Loosely install the main caps with the bolts (or studs and nuts) and tap the crank forward with a couple good whacks with a rubber hammer to set the thrust surface on the rear main. Torque the rear main caps, starting with the thrust-bearing cap (the rear main) first, then follow the torque sequence, and torque the hardware evenly in 20-pound increments (20, 40, 60, etc.) until they are all properly torqued.

ENGINE REBUILDING

 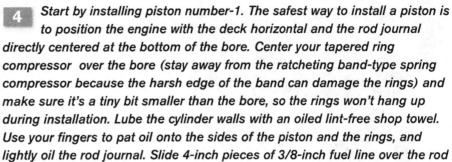

4 *Start by installing piston number-1. The safest way to install a piston is to position the engine with the deck horizontal and the rod journal directly centered at the bottom of the bore. Center your tapered ring compressor over the bore (stay away from the ratcheting band-type spring compressor because the harsh edge of the band can damage the rings) and make sure it's a tiny bit smaller than the bore, so the rings won't hang up during installation. Lube the cylinder walls with an oiled lint-free shop towel. Use your fingers to pat oil onto the sides of the piston and the rings, and lightly oil the rod journal. Slide 4-inch pieces of 3/8-inch fuel line over the rod*

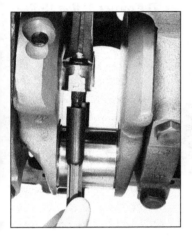

bolts to protect the crank from being nicked with the rod bolt, or simply use a Wildman Rod Guide tool for extreme protection. Make sure the rod chamfer is correctly oriented on the rod journal fillet and the piston and ring gaps are in the correct positions. Lower the piston into the ring compressor slowly to make sure the rings don't bind in the compressor (you can wiggle the piston around a little). If the ring pops out and binds the piston, raise the piston and start again. Push the piston with your thumbs. If that isn't enough and you're sure the rod isn't caught on the crankshaft and the rings aren't bound up, tap the top of the piston with a dead-blow hammer. If you tap too hard, the bearing will pop out of the rod and smash on the ground. It should not take more than a couple light taps. Confirm that the rod seats against the crank properly and that the bearing is still installed. Evenly tighten the rod cap (with bearing installed), but don't torque it yet. Install piston numbers-3, -5, and -7 using the same methods, then rotate the engine on the stand so the even bank of cylinders is horizontal and repeat the process.

The lubricant you put on the camshaft is extremely important, especially for flat-tappet camshafts. Only two lubricants should be used on the extremely high-pressure contact surface of a camshaft. The old standard is liquid moly lubricant, not spray-on moly lube or bolt-thread assembly lubricant. Only use moly lube made specifically for lubricating camshafts. Joe Gibbs Driven Assembly Grease is race-proven to be tough enough to take the place of moly lube. All the $70,000 NASCAR racing engines have flat-tappet camshafts, and Joe Gibbs Racing uses Joe Gibbs Driven Assembly Grease for all of its engine builds. Of course using the proper lubricant is not the only part of keeping your camshaft from going flat—there's also the necessary break-in procedure upon start-up.

If you want to upgrade to a hassle-free hydraulic-roller camshaft, use this retrofit kit from Speed Pro. It reduces friction in the engine for more horsepower. Solid-roller lifters add maintenance and unwanted valvetrain noise. The hydraulic-roller cams combine the low maintenance of a hydraulic lifter and the reduced friction and extra long life of a roller lifter. The beauty of these cams and lifters is that the only break-in period is for checking for leaks and starting the ring-seating process. A flat-tappet cam and valvetrain is much more demanding to install and break in.

Engine Assembly CONTINUED

5 We're installing a Speed Pro double-roller timing chain and like most others, it may have some fitment issues. Early Chevrolet small-block engines were never designed to have a wide double-roller timing chain. Therefore, the blocks require a little clearancing behind the timing chain, so the high spots (shown with arrows) are removed. Grinding anything in the engine should be done while the engine is at the machine shop, not after the engine is partially assembled due to the metal shavings and debris that could make their way into the engine. These high spots aren't always a problem, but taking some time to grind them beforehand thwarts a nasty surprise later. The owner of this 1968 block was really lucky that the timing chain didn't explode when it was grinding against the big square-drive plug on the right side of the photo.

 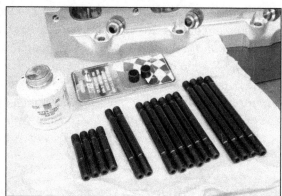

6 Before installing the heads, we installed new Mr. Gasket dowel pins. They align the head and also aid in keeping the head gasket in place. We put ARP thread sealer on the threads and all the ARP head bolts because all the bolts thread into water jackets. (Our Victor Reinz head gaskets have the words "up" and "front" so we know how to install them correctly.) Make one last inspection for loose debris inside all the head ports prior to installation—one little nut, screw, or rag can ruin your day. When you're ready to lay the head on the block, place two head bolts on a rag in the lifter valley to easily grab and install them in the block. This prevents the head from sliding off the engine and maiming you and the head. Make sure the head is completely seated on the block and dowel pins before torquing a bolt. Torque the bolts in 20 ft-lbs increments in the correct sequence. The torque spec changes when installing ARP bolts, so check the paperwork that comes with each bolt set. The spec also changes depending upon the thread lubricant you use (ARP Thread Sealer, assembly lubricant, or 30W oil). Factory torque specs are calculated using the equivalent of 30W engine oil as a thread lubricant.

7 We installed our heads with new hardened valve seats and Speed Pro stainless-steel valves. To increase performance, Bob Gromm installed the next-size-up exhaust valve and corresponding seat. We chose the Speed Pro retrofit hydraulic-roller cam kit, which included the necessary shorter pushrods (the lifters are taller than flat-tappet versions), valvesprings, locks, and retainers. We installed them in the heads with the correct valve height by using a Goodson valvespring height gauge and installing the correct number of hardened shims.

ENGINE REBUILDING

 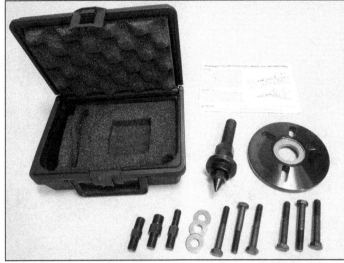

8 *The elastomer band between the hub and outer ring on our original harmonic damper had deteriorated. We compared it to another damper, and our hunch was correct—the outer ring had moved about 1/4 inch from its original location. When this happens, you never get an accurate reading from the damper's timing mark. If you've ever set the timing on your car and it runs poorly afterward, it's possible your outer ring has spun. Once the damper starts to spin, the ring can completely come off the hub in extreme conditions. Since we want to reuse our original damper, we had Damper Doctors rebuild it. It was disassembled, cleaned, and inspected for cracks (which are common around the keyway). The timing mark clocking was set, and it was injected with high-heat and high-strength silicone that was platinum cured for additional strength. There are dampers with various diameters and thicknesses for different applications. If your damper is original, you can have it rebuilt or install the correct reproduction unit (remember, your timing tab may not work with the wrong damper).*

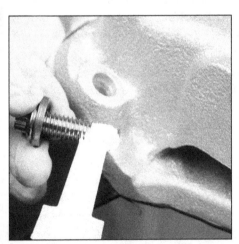  

9 *Test fit the intake manifold to the block and heads to make sure the cylinder-head angles exactly match those of the intake manifold. If they don't, you'll have all sorts of coolant, vacuum, and oil leaks. Depending on the incorrect angles, you could have these leaks into the lifter valley and not even see them. A shop can machine the intake manifold to obtain the correct angles. All small- and big-block intake-manifold gasket kits come with cork and/or rubber end seals that fit between the block and intake. Historically, these pop out and create a huge oil leak, so most engine builders toss them out and install a 1/4-inch-diameter bead of Permatex Ultra Black silicone gasket maker across the front and back rails of the block. If you're going for a factory look, install the rubber end seals and take your chances.*

Engine Assembly CONTINUED

10 In order to install the distributor correctly, the tang in the center of the distributor gear must engage the slot in the oil-pump driveshaft. To move the shaft into proper position, insert a long standard screwdriver into the hole in the intake manifold. Keep in mind that the gear has helical-cut teeth and when installing the distributor, these teeth mesh with the teeth on the cam causing the distributor shaft to turn, so you have to compensate for that movement. When the distributor is fully installed against the intake manifold, the ignition rotor tang needs to point toward cylinder number-2. To establish correct distributor height, install the distributor without the gasket to see if it completely seats against the intake manifold. If it sits up off the manifold, get a Mr. Gasket distributor shim kit (#6059), which has .030-, .060-, and .090-inch-thickness shims. Measure the gap and add .030 to the measurement. Place the shim and the gasket between the distributor and the intake manifold. Be sure to set the correct distributor height; otherwise you'll blow up your oil pump and other parts in a hurry. Don't forget to coat the distributor gear and the tang in the end of the distributor with some JGD assembly grease.

11 Before you fire up your engine, don't forget the most important last step: Make sure the oil drain plug is installed and tight, but not over-torqued to the point of cracking the Teflon gasket. Make sure you've installed the oil filter. If you have a 1967, it is equipped with a canister oil filter with a replaceable element. In 1968, Chevy switched to the conventional spin-on oil filter. If you're interested in a correct-looking oil filter, get an AC reproduction red, white, and blue oil filter. Oil filters are not all made the same, so be sure to buy a good-quality oil filter. If you don't care about the correctness of the oil filter, WIX makes a great one (P/N 51069 for standard and P/N 51069R for performance applications). Don't use the "R" filter for initial start-up and break-in. After break-in you can switch to the R filter. The tall one is for trucks.

ENGINE REBUILDING

A few oil companies are now offering specific oil for older cars. One of those is Joe Gibbs Driven. One of its products is BR, which is specially formulated oil with extra additives specifically for breaking in an engine. This oil replaces the very old standard of breaking in an engine with 30W oil, especially now that the 30W oils are missing everything you really need upon break-in. Once the engine has been broken in, you should switch your engine to a regular diet of Joe Gibbs Driven Hot Rod oil. It has the additives to keep your engine in good working order. It has extra protection for internal engine parts that sit for extended periods of time (a month or more), especially if you live in cold climates and store your Camaros for eight months out of the year.

Put a light film of oil on the camshaft journals. Cover the distributor gear and the four rearmost cam lobes with liberal amounts of lube. Carefully slide the cam into the block, trying to keep from knocking it into the bearings. Rest it on the two rear cam journals. Lube the next set of cam lobes, slide the cam in a little farther, lube the next set of lobes, and so on. Don't forget to lube the smaller, round front lobe, which is the fuel pump lobe.

That's the cleanest way to lube a cam. If you prefer the dirty way, lube up every cam lobe at once and carefully feed it into the block, but be sure you don't accidentally wipe the lube off the camshaft during installation.

Dampers

The damper is supposed to be a press-fit. Performance dampers are supposed to be an even tighter fit. When installing your damper, coat

When it comes to certain levels of engine restoration, there is a choice to make on bolts. Even if you're restoring the exterior of your engine, it doesn't mean all the internal bolts need to be correct. This engine has every externally visible bolt correct, but the unseen bolts in the rods and mains are the best strength and quality engine bolts on the market—ARP rod bolts and main studs. The main studs are stronger and more reliable than bolts, because when you tighten a bolt, you're torquing it and grinding the threads in the block. The studs are hand threaded into the block. You do not grind the threads into the block when you torque the nut on the stud; the stud is stationary in the block. This also ensures a correct torque reading and eliminates main-cap walk (movement) under load. This engine features an oil-controlling windage tray, hence the longer main stud. There was rotating assembly interference with some of the rod bolt heads, so some had to be clearanced.

When the engine has been completely assembled, it will look completely stock on the outside, even though the internal parts were replaced with upgraded parts for higher performance. When this small-block is complete, with valve covers over the roller rockers, a stock-looking carburetor, cast-iron exhaust manifolds, Chevy Orange paint, and all the accessories, nobody will know it's not completely stock, except the driver.

the crank snout and bore of the damper with Permatex anti-seize. Put a thin film of oil on the damper seal and damper snout. Do not use a hammer to smash the damper onto the crank; it damages the damper and the crank thrust bearing. Don't try to draw the damper onto the crank with a long bolt. Tightening a bolt into the crank pulls the threads. Use a damper installation tool, the correct tool for the job, to pull it onto the crank.

Gaskets

After years of operation, old heads and intake manifolds have rust pitting on the gasket surfaces surrounding the coolant passages. To help seal this area we use a thin film of Permatex Water Pump & Thermostat Housing silicone that's made specifically to resist antifreeze. Before choosing your gaskets, keep in mind that aluminum intake manifolds work best with composite gaskets and cast-iron intake manifolds work best with steel-core gaskets. We chose Victor steel-core gaskets for our small-block project.

Picking Parts

Purists rebuild an engine to original factory specs, and they may choose a camshaft that matches the original cam specs, compression ratio, pistons, performance rods, steel crankshaft, etc. This is an effort to build the engine back to all-original horsepower specifications. Truthfully, even restorers you might think would go that route, don't. The reason is that if everything looks correct on the outside, who's really going to know what the inside of the engine looks like? Nobody is ever going to know if you have installed Eagle crankshaft and rods, higher-compression pistons, a hydraulic roller high-lift camshaft, undercut stainless-steel valves, roller rocker arms, etc. (Be aware that some rocker arms require tall posi-locking rocker nuts that won't fit under stock valve covers without some extra machining to the nut and/or the rocker arm stud to reduce the overall height). If you can hide it under the valve covers or inside the engine it won't matter to anyone else

An efficient and appropriately matched exhaust system is essential for any high-performance engine build, so installing a high-performance head, cam, and valvetrain will be all for nothing if you don't have free-flowing exhaust. Factory performance Camaros could be ordered with headers, which came in the trunk. It was up to the customer to install them or have the dealership do the job.

There are multiple sources for aftermarket parts. If you're in the market for a high-powered and high-revving engine, take a good look at what you're really going to do with the car. It's not a bad thing to over-build the engine, but do it within reason and build it to meet your realistic driving requirements. Don't make silly mistakes by underestimating and matching weak parts with high-performance parts, such as building a 600-hp engine with a stock cast-iron crankshaft.

If you decide to rebuild your engine to be closer to the factory output and fairly mild, you can save some money by purchasing separate parts or a kit from Sealed Power. For our small-block project engine, we chose Victor gaskets, Clevite bearings, and Sealed Power cam, lifter kit, and oil pump. The small-block has a factory-forged crankshaft.

If you're dead set on looking original on the outside and you're keeping the cast-iron exhaust manifolds, you can do what this enthusiast did. Use the extrude honing process to port out the exhaust manifolds for additional exhaust flow without installing headers. Cast-iron and aluminum intake manifolds can also benefit from this process.

ENGINE REBUILDING

These heater hoses have the correct ribs and clamps for the build date of this Yenko. The bypass hose, edges of the aluminum intake, valve cover lips, and gasket have overspray because these parts were installed when the engine was painted at the Tonawanda engine assembly plant. A masking cover was placed over parts of the engine to keep the overspray to a minimum.

The exhaust manifolds were installed on the big-block engines before Chevy Orange was applied to the engine, so exhaust manifolds got overspray around their bolts and bead-bolt flanges. Smog equipment, sensors, and other accessories were added after paint. Each engine line painting procedure was different, so to get factory-correct finishes, research your specific application.

If you run into original clamps here's how to loosen them: Loosen the bolt and press down the top of the clamp, and the clamp will come loose. Save your old ones and re-use them or sell them to a restorer. The original clamps have specific markings; reproduction clamps are just not the same.

CHAPTER 5

Extrude Honing

Extrude honing components, such as intake manifolds and exhaust manifolds, increase engine power output. I've seen tests of an aftermarket aluminum dual-plane intake manifold with this process and the horsepower and torque increases were in the single digits (1 to 9), which isn't much. If you're stuck with stock parts, however, it may be worth it and there could be a larger gain with a stock intake manifold. Just be careful not to hone the exhaust manifolds too much and make them weak to where they crack easily after normal use.

Engine Fasteners

If you're using factory-correct parts, hopefully the engine you're rebuilding is correct for your car and still has its original engine fasteners. If you need to source new bolts, you can find some new reproduction bolts at Year One, Classic Industries, and other reproduction parts suppliers. AMK Products and Paragon Corvette Reproductions also sell correct bolt kits that match the originals from the factory.

At a minimum you should upgrade the fasteners that are hidden inside the engine. Even the prying eyes of judges are not able to see them if you're building your car to be factory correct. We don't like to reuse original main bolts or rod bolts, so we always upgrade to ARP fastener rod-bolts and either ARP or Milodon main bolts or studs on the bottom end. High-quality fasteners are important insurance and provide reliable clamping force. These are definitely worth the expense to avoid a potential engine failure.

If you're rebuilding a stock engine from 40 years ago, the stock bolts have been heat cycled so many times—and have probably been over-stretched—that you should not reuse most of the critical fasteners. ARP bolts are the best bolts on the market for critical and non-critical engine fasteners. Its bolts are rated much stronger than Grade 8 bolts in their yield and tensile strength. Its threads are more precise than most other bolts, which helps protect the threads in your parts for years to come.

ARP bolt heads and studs stick up higher than the low-profile factory bolt heads. They interfere with some exhaust manifolds and headers, so check the clearance or you may burn through a couple of gaskets before you notice the issue. If you're not worried about the non-factory bolts, but want bolts that won't ever turn ugly and rust after they've been in use for a few thousand miles, spend a few extra bucks and install ARP stainless-steel fasteners.

Factory Engine Detailing

Sure you may know how to detail your own engine the way you want it, but here's a little info about how the engine that left the factory would have looked.

The factory painted the engines when they were fairly complete. In fact, the engines were painted when the intake manifold, valve covers, exhaust manifolds (on big-blocks), water pump, and oil pan were already installed. In addition, the big-block Chevy engine had the short water-bypass hose installed when it was painted, so when you see the hose, lower edges of chrome valve covers, and edges of aluminum intake manifolds painted Chevy orange on a restored car, you know the builder wasn't paying attention to this detail because it was done that way on purpose. Maybe it doesn't look as good as the way you would have it, but it is correct.

Keep in mind that the engine-assembling facility didn't have hours on end to mask everything off or remove parts in order to paint them neatly. In fact it's been speculated that the Tonawanda plant used covers simply laid over the aluminum intake manifold and chrome valve covers. Some people go overboard with the overspray on the intake manifold to where there's more than a 1/4 inch overspray area around the head sealing surface.

Factory Clamps

Most Camaros have been worked on over the years, so a lot of the small parts have been removed and replaced. For some reason auto mechanics—and even hobbyists—feel it is necessary to remove the factory tower-style hose clamps and replace them with worm-gear clamps, but these don't seal well because they pull the hose tighter around the gear area. Factory clamps actually sealed more evenly around the whole clamp.

In order to remove tower-style clamps, you loosen the bolt, but don't remove it, and push down on the top of the tower. Don't push down on the bolt because you'll push the captured nut out of the clamp and bend or break the tab that holds the nut in place. Only push down on the top of the tower. If you have original tower-style clamps and are not going to use them, you may be able to sell the originals to restorers.

CHAPTER 6

TRANSMISSIONS

Start by draining as much fluid as possible from the transmission. Some cases have a fill plug and a drain plug and some only have a fill plug. Fluid gets trapped in the tailshaft housing, so expect additional fluid to need to be drained. You won't get all the fluid out, so disassemble the transmission on cardboard to soak up all the leftover fluid. This also prevents the case from being damaged on the metal workbench.

The following info about power ratings and applications are based on factory application data and common knowledge about the strength of the different transmissions. We also assume that even most of the enthusiasts who are restoring Camaros to the most finite detail want to upgrade the internal components of their engines—this produces more horsepower and torque than the original transmission was designed to handle. As a result, your Camaro must be equipped with a transmission that can reliably transmit the higher torque levels. Therefore, we cover higher-performance parts and upgrades, as well as some brief information on overdrive transmissions.

There are entire books covering transmission rebuilding. One of them is published by CarTech, *How to Rebuild and Modify High-Performance Manual Transmissions* by Paul Cangialosi. It has some great tips. In this chapter we cover as much detail as we can.

Removal Safety

When removing a manual or automatic transmission from your first-generation Camaro, always start with safety steps. We're going to assume you don't have a lift or a hoist and will be working on your back under your Camaro. The first thing you should do is park your car on a solid, level surface. Place large blocks of wood or actual wheel chocks behind and in front of both rear tires. If you fail to block the wheels, your car could roll off the jack stands. Block both rear tires so the car can't roll forward or backward. You can't rely on the park sprag in the automatic transmission, the manual transmission, or the driveshaft to keep your car from rolling because you're removing these parts! Don't trust the parking brake either.

CHAPTER 6

Safely jack up the front of your vehicle with sturdy jack stands to keep your car from falling on you when you're moving the car around during transmission removal and installation. Nothing is safer than using jack stands. Never get under a car that is supported only by a floor jack. You could get crushed to death if it fails.

To remove the driveshaft, you have to remove the U-joint straps from the pinion on the differential. When you remove the driveshaft, the suspension unloads if there's any stress on it. The driveshaft weighs only about 10 pounds. Before you pull the yoke out of the back of the transmission get an oil-changing pan to catch all the fluid that pours out. Do not get fluid on yourself, especially if it's hot. The worst smelling fluid to get on you is old manual transmission or differential gear oil. You've been warned!

Before removing the automatic transmission from the engine, you need to remove the three bolts holding the torque converter to the flex plate then push the converter into the transmission about 1/4 inch. When removing the automatic transmission, make sure the torque converter doesn't slide off the front of the transmission on its own and surprise you. Either remove it or get a metal strap to go across the front of it to keep it in place.

Unless you're a really strong person, you're going to need a transmission jack to remove an automatic transmission. Even if you are strong, the transmission could crush you or maim you if it fell on your chest or on a finger, so be very careful and safe. The Saginaw is pretty hefty with its cast-iron case, but the Muncies are a little lighter. All transmissions weigh less when the fluid is removed. A manual-transmission flywheel can weigh as much as 60 pounds if it's steel, so when you remove the last bolt, be prepared.

Manual Transmissions

We cover 4-speed transmissions in this book because most restorers

We picked up these Chevrolet Chassis Overhaul Manuals *from Year One. These manuals show how to rebuild all the Camaro manual and automatic transmissions offered the specific year of the manual. They only cover rebuilding to factory specs and have limited black-and-white photos/illustrations. They aren't extremely intuitive, but they're very helpful.*

are not going to be working on 3-speed manual transmissions. If you are working on a 3-speed, pick up a Chevrolet *Chassis Overhaul Manual*. The most common manual-transmission rebuilds for first-generation Camaros are 4-speeds from Saginaw (RPO M20) and Muncie (RPO M20, 21, and 22).

Physical Attributes for Identification

Here are a few identifying features to help distinguish between the Saginaw and Muncie 4-speeds without having to look inside them.

Saginaw

The main case on a Saginaw transmission is cast iron, which adds a considerable amount of weight compared to a Muncie. The Saginaw is also easily distinguishable because it has the reverse gear selector in the side cover next to the first- and second-gear selector. Because the transmission doesn't have a separate rear bearing retaining plate, like the Muncie, the tailshaft housing bolts directly to the transmission with five bolts.

Muncie

If it's an original 4-speed with aluminum case in a first-generation Camaro, it's a Muncie. The reverse gear selector and reverse gear assembly are located in the tailshaft housing. The Muncie also has a rear-bearing retainer plate sandwiched between the main case and the tailshaft housing. The 1963 through 1968 Muncies have a stud sticking out of the gear selector shafts, so the shifter linkage plates are attached to the 1-2 and 3-4 gear and reverse selectors with a nut. The 1969-and-later Muncies had a tapped hole in the gear selector shaft, so the linkage plates attached with bolts. In general, other than those external differences, the Muncie 20, 21, and 22 have the same type of overall design.

TRANSMISSIONS

Yes, you just read that correctly: Saginaw and Muncie both produced 4-speeds for the RPO M20–equipped Camaros. The Saginaw M20 was a weaker wide-ratio transmission used in the L6 (6-cylinder) and 210-hp V-8 models. The Muncie M20 was a stronger wide-ratio unit used in Camaros with the RPO that were equipped with 325-or-less-hp V-8s. The Muncie M21 and M22 transmissions were close-ratio and were used in many different horsepower levels of V-8-equipped Camaros. Because we're restoring and modifying first-generations to the higher performance level, the focus of most of the information in this chapter is on the M21 and M22 transmissions.

The factory *Chassis Overhaul Manuals* provide instruction on rebuilding all the automatic and manual transmissions offered by the factory (as well as the rest of the chassis components). They cover all the specialty tools needed in a perfect world. Unfortunately, many of the tools are no longer available and you need to be industrious to figure out how to make due with substitutions you may have in your own toolbox. Don't get us wrong, these manuals are awesome. You just need to exercise your perception and reasoning skills as a mechanic to follow some of the instructions. The exploded views of the components are invaluable and make rebuilding so much easier. Using this book and the Chevrolet factory manuals helps overcome obstacles while rebuilding a transmission.

4-Speed Manufacturers

The Saginaw M20 and Muncie M21/M22 were the three 4-speed transmissions offered for 1967–1969 Camaros. The Saginaw should be considered a light-duty manual and should not be mated to an engine with more than 300 hp. The famed Muncie M21 and M22 "Rock Crusher" transmissions are tough and reliable manual transmissions that have served the Camaro well. In fact, the Rock Crusher was fitted to the 1967–1969 Z28s as well as the big-block Camaros.

Borg-Warner

The Borg-Warner 4-speed transmission was never factory equipment for the first-generation Camaro. The 3-speeds were the only Borg-Warner transmissions installed in 1967–1969 Camaros. (We won't be mentioning the T10, even though it was installed in some Camaros after they left the factory, including cars modified at Yenko.)

Saginaw

The Saginaw M20 4-speed transmission is best suited for engines up to 300 hp (even though Chevrolet only offered them in Camaros with less than 210 hp) because when you put more power in front of a Saginaw, the gears are smaller and weaker they lose more teeth than a cowboy in a bar fight.

Muncie

A Muncie M22 was selected for covering our rebuild process. The Muncie M21 and M22 transmissions are the most popular high-performance manual transmissions for the Camaro. Of these two, the M21 is more common, while the M22 is a tougher, higher-performance transmission known as the legendary "Rock Crusher" because of its somewhat-straight-cut gears. Well, if you didn't know this, your bubble is about to burst. The M22 gears are not actually straight-cut; rather, they are helical-cut, but they are more straight-cut than the M21 gear. The M22's gears create a louder, distinctive gear whine than other production transmissions, but the whine would be unbearable for the typical street-driven performance Camaro. Straight-cut gears are reserved for full-race transmissions and in some cases those gears can be heard over an uncorked racing engine.

The M22 is known as the "Rock Crusher" because of its straight-cut gears. As you see here, the M22 gear (on the left) is compared to the M21 gear on the right. The M22 gear is not straight-cut, but it is cut at less of an angle. You can also see that the teeth are much more robust on the M22 gear, compared to the M21 gear.

There were many different tailshafts with different configurations. The transmission we're rebuilding has a Chevelle tailshaft, but the one shown here has the correct shifter location holes for a first-generation Camaro.

CHAPTER 6

With the exception of typical hand tools and a lead hammer, specialty tools are required to rebuild a Muncie M21 or M22 because they are very similar in every aspect. From left to right, these specialty tools are: Snap-On (#SRP2 or new #SRP2A) snap-ring pliers with externally knurled jaws and dimples (for assisting in removing snap rings with pointed ends), a generic gasket scraper with heavy-duty blade, a bearing and seal driver set, a Muncie bearing retainer nut wrench, a special wrench Mark Schwartz made out of an old wheel weight (for holding the reverse-gear selector spring-loaded ball-bearing for installation), and a ball-peen hammer...obviously. The first order of business is to remove the four bolts from the front shaft support and slide it off.

Muncie M22 Rebuild

1 Shift the transmission into two gears at once to lock the shaft is in place. Remove the main-drive-bearing retainer nut from the output shaft, which has two flat sides on the back side. A special wrench is needed to remove the retainer nut. Some inexperienced people have used a pipe wrench to remove it, and damaged it, like the one on this transmission. The flange is damaged with tool marks, which causes front shaft leaks. The nut and main-gear shaft have left-handed threads instead of typical right-handed threads.

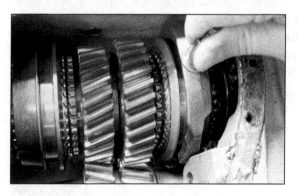

2 Before removing the side cover, shift the transmission into second gear. If you don't have the shifter handy in order to shift into second gear, use a large Crescent wrench to turn the shifter shaft on the right side of the cover counter-clockwise from a straight up-and-down position. Doing this moves the shift fork forward, so you can remove the cover without binding the shift fork against the case. If you don't, the cover will bind a little and the fork will pop out of the shifter shaft. It's not the end of the world if it falls off, we're just mentioning it to make life easier.

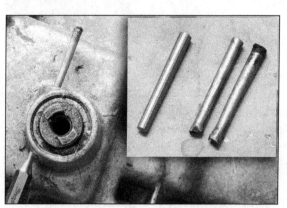

3 There's a tapered pin in the reverse-shift shaft that must be removed and is easier to do so while the tailshaft housing is still attached to the transmission. The pin is tapered and is installed from the top. This way, if it comes loose, gravity keeps it in position. To remove it, drive it out from the bottom with a hammer and punch. The inset photo shows two old pins that are bulged from being driven out and reused too many times. Some rebuild kits contain new pins. If you damage your pins when using a drift punch, you may need to get a new one separately. Don't use a mushroomed drive pin in the housing or you'll damage the housing. Some kits supply a roll pin as a replacement. Shift the shifter shaft into reverse as shown.

Muncie M22 Rebuild CONTINUED

4 Before pulling the case apart, pull the reverse gear shifter shaft out until it stops. This helps disengage the reverse-gear shifting fork from the reverse gear. Remove the six bolts attaching the tailshaft housing to the main case. Once all the bolts are out, separate the tailshaft housing from the case and the plate sandwiched between the two called a rear-bearing retainer plate. There's a dowel pin in the bearing retainer keeping everything from twisting apart. This transmission case took a little coaxing with a dead-blow hammer and when there was a small gap, we were able to insert our wide flat-blade gasket scraper between them to separate everything. Don't jamb a screwdriver between the case parts or you'll damage the gasket sealing surfaces.

5 Remove the reverse-idler gear assembly from the tailshaft housing, then the reverse-gear shifter shaft and shift fork, but be careful because there's a spring-loaded ball under it and you don't want to lose the ball or the spring. Remove the rear yoke seal with a wide chisel. Then place the housing on a flat wooden surface and drive the rear case bushing sleeve. Drive it out by pounding it into the housing. Line your narrow chisel up with the slot (highlighted), so you don't damage the inside of the housing. Start pounding the bushing and drive it into the housing until it's all the way out.

6 Pull the mainshaft out of the case. The rear-bearing retainer plate is pressed onto the mainshaft, so it and the gear cluster come out in one assembly. Set these aside. Remove any errant parts from inside the case. Rotate the case up on the rear gasket surface and place it on a flat wooden surface to protect the gasket area. This image shows a hand holding the assembly by the shaft, but a free hand should be holding the main-drive gear through the sidecover hole, so it doesn't fall to the floor and get damaged. Use a lead hammer for this step or else you'll damage the shaft. You may be able to get away with using a dead-blow hammer, but the sharp edges of the shaft may damage polyurethane striking surface. Whack the input shaft with a lead hammer until it comes out of the bearing. Keep it from falling.

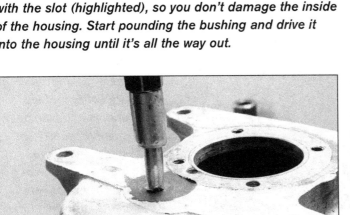

7 With the front of the transmission case still facing up and hanging off the edge of the block of wood, drive the countershaft out of the case with a hammer and big brass drift. The dark spot on the face is some silicone the previous builder used to seal the countershaft because this case is loose from many hard years of road racing. This shaft nearly fell out, which means that this case should be replaced. The shaft has to come out this direction because there's a small step on the other end of the shaft.

Muncie M22 Rebuild CONTINUED

The countershaft goes through the countergear, which rides on 112 equal-size roller bearings. These roller bearings are all over the place inside the transmission; collect them all and keep them in one area. This whole countergear assembly consists of the countergear, countershaft, roller-bearing spacer sleeve, two thrust washers, six roller-spacer washers, and the 112 roller bearings.

Place all the parts aside to be cleaned. Set them in an order to help you remember which parts go where instead of grouping them together as we have done here. When you've rebuilt numerous transmissions you'll know where the parts go without keeping them in order. Keep the different roller bearings separate from the others by containing them in the center of gears or use a couple of small containers. It's good to place rolling parts in the center of the parts to keep things organized and also to keep them from rolling away.

9 The rear bearing retainer plate and the first-gear sleeve are pressed onto the mainshaft, so they must be pressed off. First remove the snap ring, then the bearing retainer, then pull off the first gear, synchro, and sliding sleeve. Now the first- and second-gear clutch hub need to be pressed off to remove the pressed-on first-gear sleeve. Then the transmission is ready to have all the parts cleaned and evaluated to see what needs to be replaced and what upgrades you may want to consider.

8 Use snap-ring pliers to remove the gears and synchro rings. The dimples in the external faces help retain the pointed tips of the snap rings. Start by removing the front snap ring and pulling all the front parts of the mainshaft. Keep all the parts in order as they come off the shaft.

The synchros ride on the tapered surface of the gear. The synchros are softer material and wear down, but small grooves from serrations in the internal surface of the synchro wear very minor grooves in the face of the gears. If you can catch your finger nail or feel the grooves you may want to have a machine shop resurface them for you, or purchase a new gear.

TRANSMISSIONS

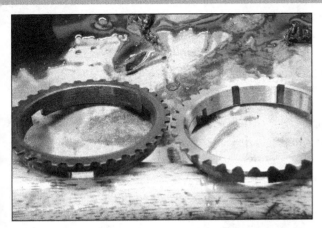

The synchro on the left is the old one. As they wear, the raised serrations on the internal surface becomes smooth. If they are smooth, it's time to replace them. The synchro on the right is a new one and features grooves cut in the serrated internal face for higher performance. When the synchro is engaged on the tapered cone of the gear, the gear oil is forced out from between the two surfaces. Oil takes longer to escape from the non-grooved synchro, and the grooved synchro engages faster because the oil has a path to escape through, which makes for faster shifting action.

10 You can make some upgrades to the transmissions, but these are usually reserved for performance gearboxes. The first gear rides on the gear oil between it and the sleeve (shown). This sleeve has had a couple of grooves machined in at an angle to get extra lubricant between the sleeve and gear. Don't groove your sleeve without professional instruction. Racing builders replace the sleeve with two-caged roller-bearing assemblies. Replace the case with a Super Case and the rear-bearing retainer plate with a stronger cast-iron unit.

11 Installing all 112 of the roller bearings back into the countergear is the hardest part of rebuilding a Muncie. You need a shaft the size of the countershaft and the length of the countergear to slide into the countergear and start installing the sleeve, washers, and rollers bearings one section at a time. Or you can simply assemble it as the professionals do it—without the tool but with white grease to keep the bearings in place. Here are a couple of bearings placed on the shaft between the spacer washers. Racing transmissions replace all that hassle with these caged bearing units and the end spacers are replaced with extra friction reducing needle bearing thrust washers that ride against the tanged washers. All these roller-bearing parts not only reduce build time and frustration, but they also greatly decrease friction, which makes for a better operating transmission.

This washer is typically found in higher-performance transmissions and rides at the base of the main drive-gear and rides against the main drive-gear bearing in the front of the transmission case.

HOW TO RESTORE YOUR CAMARO 1967–1969

Muncie M22 Rebuild CONTINUED

12 *Inspect all the parts for wear and replace damaged or worn parts. The synchros are softer material and the internal serrations wear down. The teeth on the outer edge of the synchro should look like a box with a peak. These lose their defined peaks and get flat on one side of the peaks. The tapered face on the gear where the synchro rides should not have any grooves where the synchro rides. The gears and synchros should not have any chipped or extremely worn teeth. They may be worn, but if they are missing or have no peaks they should be replaced. The main teeth on the gears and the countergear should all be intact and in good condition. The clutch keys have little nubs that ride in a small groove on the internal diameter of the clutch sliding sleeve (the ring operated by the shifting fork in the side cover). If the nubs are worn down or the grooves are worn into the faces of the keys, you should replace them. The clutch-key springs (in the shape of a C) hold the clutch keys, and these springs should all be in good condition. The mainshaft and maingear bearings should all spin by hand without having any flat spots or slight binding. The little roller bearings should all be in good shape and the countershaft should not have any grooves or steps (other than the one small step about a 1/2 inch from the end of the shaft with the large notch).*

13 *All 112 roller bearings install two ways. You can get yourself a smooth, hardened shaft with the same outside diameter as the countershaft, but the same length as the countergear to install all the bearings. We chose to use heavy white grease to hold them in place instead. Assemble everything (two stacks of 28 rollers each separated with washers and then a sleeve and then two more stacks of 28 roller bearings separated by more washers) with the white assembly grease. Lay the case horizontally on the bench. Install the front and rear thrust washers in the case with the tangs facing the case—you can do it the easier way and Super Glue them in place, but center them over the holes or they pop loose when installing the shaft. The glue only holds them long enough to slide the shaft though. Insert the counter-gear assembly between the thrust washers. The back of the countershaft has a step in it—the taller step goes upwards and the lower step is flush with the case when the shaft is completely installed. With the countershaft steps properly positioned, slide the countershaft into the case from the rear and slowly push it through the center of the countergear. As the countershaft slides in, the tool (if you're using one) should start sliding out of the front of the case if you have everything lined up and nothing is binding.*

The second method of assembly goes as follows. By hand, assemble all the parts in the center of the countergear without the countershaft tool inside. Use liberal amounts of white grease to keep bearings in place, but not so much that inserting the shaft drags the bearings out of place. Use the same method of installing the thrust washers and the countergear assembly. This is not easy. If something binds, wiggle it to see if it unbinds without popping a roller bearing out of place. If everything goes smoothly, confirm none of the roller bearings fell out of place once the shaft stoped. If they did you'll have to start this arduous process all over again. Don't be surprised if it doesn't go in correctly the first time or two. The short step in the shaft is supposed to be on the bottom of the case. Use your brass drift to sink the shaft until the short step is flush with the case, which also means the front of the shaft is flush with the front of the case.

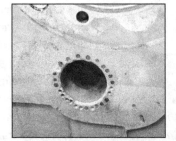

The countershaft causes problems from increased power output and severe driving. Frequent rebuilds after every few races don't help matters either. The countershaft bore in the front of the main case starts to enlarge and no longer hold the shaft in place. An enlarged countershaft-bore causes problem for the rear-bearing retainer plate because the step in the shaft is supposed to be kept stationary by the plate. When the countershaft starts to get loose, it can damage the rear plate. Someone dimpled this to reduce the hole diameter.

 **14** Both clutch hubs are the same, but the sliders are specific for first and second and different for third and fourth gears. Each synchro is specific for each gear. Grab a hub, two springs, three keys, and first and second slider. Slip the hub into the slider with the hub is grooved face on the same side as the slot for the shift fork rides. Slide the three keys into the hub slots with the face with the centered nub riding against the inside of the slider. Next, install the two springs on either side holding the keys outward against the slider. Grab the other hub, two springs, three keys, and the third and fourth sliders. Slip the clutch hub into the slider ring opposite from the other assembly (the grooved face of the clutch hub on the opposite side of the fork groove in the slider ring). Install the keys and springs.

 **15** Assemble the rear of the mainshaft first. Place the second gear over the shaft with the synchro-cone surface facing the rear of the shaft. Install the second-gear synchro ring onto the grooved side of the first and second clutch hub assembly and make sure the slots fit the keys correctly and install the first-gear synchro on the opposite side to the clutch-hub assembly. Slide the assembly over the mainshaft with the shift-fork groove on the slider facing second gear.

 **16** Slide the first and second spacer sleeve over the mainshaft and press it on using a 3/4-inch inside-diameter pipe long enough to drive the sleeve on without damaging the end of the shaft. Slide first gear onto the shaft with the synchro cone face toward the synchro, then press the rear bearing onto the shaft behind first gear with the snap-ring groove closer to first gear. Do this using a pipe with 1⅝-inch inside diameter long enough to install the gear without damaging the end of the shaft. Install the snap ring on the mainshaft directly behind the bearing. There should not be more than .005-inch gap between the bearing and the snap ring.

 **17** Slide third gear over the front of the shaft with the synchro-cone surface facing the front of the shaft. Slide the third-gear synchro into the third- and fourth-gear clutch hub assembly on the side with the shift-for groove in the sliding ring. Make sure it lines up with the keys. Slide the whole assembly over the shaft with the shift-fork groove on the third-gear side. Install the proper snap ring to hold the hub assembly in place with the ends in the spline teeth, rather than hanging in the open grooves.

 **18** Install the rear-bearing retainer plate over the rear bearing while expanding the snap ring. Slide the retainer plate until the snap ring locks in place over the entire rear bearing in its proper groove. Install rear snap ring directly behind bearing.

Muncie M22 Rebuild CONTINUED

19 Slide the anti-lash plate over the main-gear shaft if your transmission had one. Press the front bearing onto the main gear, making sure the snap-ring groove is forward, farthest away from the gear. Position the transmission with the front of the case facing down on the bench, so you can lay the main gear (front input shaft) protruding through the front of the case. Use a C-clamp (or two) on the mounting flanges if necessary to keep it in position. Install the bearing cage and the 17 roller bearings into the back of the main gear, and use the white grease to hold them in place. Install main-drive gear through the side of the case with the shaft sticking through the front bearing hole. Place the thrust bearing for the reverse idler. The tang needs to face toward the case's machined surface. Hold it in place with some Super Glue and then position the reverse-gear idler with the gear side on the face of the thrust washer. Install the fourth-gear synchro over the machined cone face of the gear with the teeth toward the teeth on the gear.

20 Position the third and fourth synchronizing sleeve forward into fourth-gear position before sliding gear assembly into the case. This helps to align the assembly with the main gear. Add a thin coat of anerobic gasket sealer and install the gasket on the forward face of the rear-bearing retaining plate to hold it in place during assembly. Lower the mainshaft assembly down into the case, while aligning the fourth-gear synchro ring with the clutch hub and clutch keys. In addition, the main gear should mesh with the countergear, and the front bearing should be aligned in its hole in the front of the case.

21 Line up the dowel pin in the bearing retainer plate. Use a soft-faced plastic hammer to tap the bearing retainer plate to get it to seat against the case. If it binds, stop and determine where it's hung up. The roller bearings inside the main gear may have come out and kept the assembly from going in the last 1/2 inch or so. When the bearing retainer is completely seated, slide the rearmost reverse idler gear into the center of the forward-reverse gear in the back of the case and fully engage the splines. Place the thrust washer on the top of the reverse gear and slide idler shaft through the center of the thrust washer and reverse-idler gears to fully engage it in the case. Position the roll pin in line with the top and bottom of case, or the tailshaft housing will not fit.

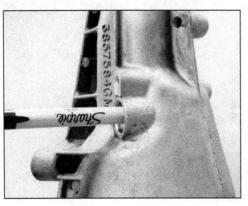

22 Place the tailshaft housing on the transmission and seat it all the way down. Use a Sharpie to mark the location of the speedometer gear on the shaft, but make sure you mark the shaft in the center of the hole. Now install the reverse gear onto the mainshaft with the flanged side facing away from the transmission. The metal speedometer-gears are soft and can gall the hardened shaft. You can heat up the metal speedometer gear to slide it over the shaft with a pair of pliers. Don't forget to take the width of the gear into consideration and center it on your mark.

TRANSMISSIONS

23 Place the tailshaft housing face down on a flat wooden surface. Using a bearing-race installation tool or the proper installation tool (shown), drive the bushing into the housing and stop when the bushing is flush with the raised edge. Then tap the new seal on the end of the housing into place. Put a light film of anerobic gasket-sealer around the seal where the housing meets the steel collar.

24 This tool is used to compress the detent ball-and-spring in the tailshaft housing. Mark made this tool out of a lead wheel-weight many years ago, but if you have a narrow putty knife or equivalent, compress the ball in order to install the reverse shifter shaft. Grease the detent ball-and-spring and put them in place. Hold the housing level and compress the detent ball and slide the shifter shaft into place while removing the tool. Push the shaft all the way into the housing and position the leg of the shaft as far forward toward the gasket surface as possible. Slide the shifter shaft seal over the shaft and tap it on with a deep socket while holding the shaft in place. With a little grease on the reverse shift-fork stud and on the leg of the shaft, position it parallel to the housing gasket surface. Thread a nut or a bolt, depending on the year, and a washer to keep the shaft from popping out during the next few steps.

25 Put a thin film of anerobic gasket sealer on both sides of the gasket and allow them to skin slightly. Lower the tailshaft housing over the mainshaft with your right hand. With the housing about 2 inches from the transmission case, use your left-hand fingertips to hold the reverse gear up and guide the reverse shift fork over the flange on the gear and push the reverse-shifter shaft into the housing with the upper palm of your hand. If the gear is engaged properly, the reverse gear should have resistance—if you push it up and it should fall. Fully seat the tailshaft housing against the transmission and engage the dowel pin. Torque the three upper bolts to 20 ft-lbs and the lower three to 30 ft-lbs.

26 The reverse-shifter shaft should be pushed all the way in. Use your wrench or shifter plate to confirm the shifting is correct. If it is, drive the tapered reverse-locking pin into the housing from the top, with the tapered end in first. If yours is mushroomed and doesn't fit, get a new one or retaper the end before forcing it in the housing. Use a drift punch to drive it in until it's snug and put a dab of anerobic gasket sealer in the holes to keep everything sealed and in place.

CHAPTER 6

Muncie M22 Rebuild CONTINUED

27 Install snap ring on front bearing. Push the third and fourth sliding sleeve forward to engage fourth gear and do the same with the first and second sliding sleeve to shift into second. Now that the transmission is in two gears at once, you can install the mainshaft nut. Use a new nut every time you rebuild a transmission unless it's a racing transmission you plan on rebuilding often. Don't forget that the nut has left-handed threads. Install the nut with the wrenching side facing the transmission and the flanged side that seals against the bearing retainer facing forward. Torque the nut to 40 ft-lbs. Then use a center punch to stake it directly over the hole in the threads on the mainshaft. Be careful not to damage the threads on the mainshaft. This locks the nut in place so it won't come loose.

28 The nut flange faces this surface. Use a thin coat of anerobic gasket sealer on both sides of the gasket and allow them to skin. Put a little sealer on the bolt threads before you install them to help seal them. Install the front bearing retainer and torque the bolts to 25 ft-lbs. If you're using the factory bolt locking plates, lock the bolts in place.

29 Before the last step, verify that everything has been assembled correctly. Center the sliding sleeves and shift out of reverse to put the transmission into neutral. Install a yoke in the tailshaft. Squirt a little gear lube down into the gears and sliders and confirm everything is properly operating. To do this, turn the input and output shafts and shift the transmission into all four gears and reverse.

30 Shift the first and second sliding sleeve forward (in second gear), in order to install the shift forks without having them bind on the reverse gear reinforcement in the case. Put a thin coat of anerobic gasket-sealer on both sides of the cover seal. Put some grease on the shift forks and stick them in the side-cover shifter shaft assemblies. Put the front shift fork in the center and rear shift fork forward. The forks should line up with a little maneuvering. Get the dowel positioned with the main case and side cover. Torque all seven side-cover bolts to 20 ft-lbs.

Overdrive Transmissions

Even though this isn't a book for pro touring, we can't assume that a 3-speed automatic and a 4-speed manual transmission are going to be enough of a final-drive ratio for the desired fuel economy, or strong enough to withstand the engine's horsepower. So this is a quick mention of overdrive transmissions. Keisler Engineering is one of the best one-stop shops for overdrive automatics. The automatics can handle 450 to 650 ft-lbs of torque; the 5- and 6-speed manual transmissions are rated for 500 to 650 ft-lbs of torque. Check Keisler's current available transmissions to see if it has a kit that suits your application. It is adding new kits all the time.

Keisler's kits are available in Basic and PerfectFit versions. The Basic kits are budget-friendly for do-it-yourselfers. The PerfectFit kits have custom bolt-in crossmembers, upgraded kit parts, and upgraded shifters all designed to make sure the kits fit in your Camaro without cutting or with very limited trimming of the floorpan for a perfect fit. The automatic kits have shifter kits that allow you to use the original shifter with upgraded extra gear on the shifter and on the shift indicator plate.

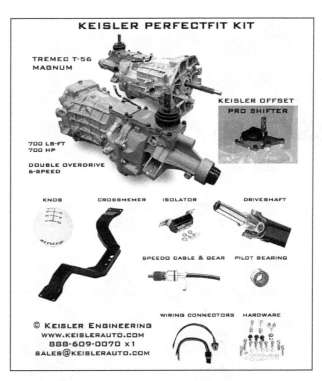

If you want a stronger automatic or manual transmission with overdrive gears to deliver better fuel economy, check out Keisler Engineering. It has unmatched quality and customer service for overdrive transmissions. (Courtesy Keisler Engineering)

Automatic Transmissions

We decided not to cover the rebuilding process of an automatic transmission because it would take about four times as many pages as a manual-transmission rebuild, and might turn into a whole book on its own.

Compared to rebuilding a TH400, the Muncie 4-speed is much simpler. You can see how everything works just by slipping off the side cover and actuating the sleeves back and forth while turning the input shaft. Pressurized by fluid, the automatic transmission operates a bunch of spring-loaded pistons and clutch packs. Because the inner workings are not simple gears, it seems as though there's a bunch of witchcraft going on inside the case, but these transmissions are not overly complex.

The *Chassis Overhaul Manual* for your car or transmission provides all the information needed to rebuild your particular automatic. In addition, it gives you a full list of specialty tools. But truthfully, we've witnessed people rebuild automatic transmissions with ordinary hand tools. Of course, these were seasoned veterans that were very industrious and knew what they could get away with and what they couldn't.

Models

The Powerglide is a 2-speed transmission and was the only automatic transmission available for L6s and small-block V-8s (except for Z28s and late 1968 TH350 testing) in 1967 and 1968.

The 3-speed transmission TH350 was an option starting in 1969. The TH350 was a huge hit with the public and outsold the Powerglide.

The TH400 transmission for 1967 and 1968 was the only automatic available for big-blocks and was only available on the 325-hp big-blocks. In 1969, the TH400 was also available for 350-hp and 375-hp big-block 396s and 427s.

Identifying Transmissions

Automatic transmission are most easily identified by the pan on the bottom. The distinct Powerglide pan is nearly square and has a small step. The TH350 pan is square with one angled corner. The TH400 pan has been described to me as resembling the shape of Michigan or Illinois.

CHAPTER 7

DIFFERENTIALS

The Chevy 12-bolt differential is one of the most revered and sought-after differentials for first-generation Camaros. In fact, a 12-bolt-equipped Camaro is worth significantly more than one that is not. Produced from 1965–1972, the differential is a strong, efficient rear-end design that effectively transmits torque to the axle shafts. The ring gear carrier measures 8.875 inches and two internal main-caps hold it in the carrier; the pinion is 1.625 inches in diameter. The 12-bolt readily withstands torque from the various Chevy small-block V-8s and supports most big-block engines up to about 900 hp. For many years, original 12-bolt rear ends were hard to come by, but now there are several aftermarket offerings if factory-original equipment is less of a priority. Strange Engineering offers new improved-design housings for first-generation Camaros.

The process of rebuilding a differential covered in this section is a detailed overview. Various specific tools are necessary to perform a complete rebuild. Some tools are a simple design, so you could make them yourself with some steel, a grinder,

One 12-bolt differential was made for trucks and one was made for cars. They are easy to tell apart. The truck differential is larger and has two (top and bottom) defined flat sections on the rear differential cover; the car 12-bolt cover is round and symmetrical. Even though both have an 8.875-inch ring gear diameter, all of their parts are exclusive. There are also 12 bolts attaching the ring gear to the carrier.

and a little ingenuity. But you may need to build some specific precision tools, such as a special fixture for a dial indicator that checks pinion gearhead depth inside the differential. If you're going to rebuild a few differentials in your lifetime, it is worth it for you to invest in some of the specialty tools. On the other hand, if you're only going to rebuild one differential, it may not make sense to learn all the procedures and invest in all the specialty tools.

Besides specific tools and techniques, rebuilding a differential requires a high-level of detail to get the correct shims, backlash settings, and other aspects of the build. While the level of precision isn't as high as rebuilding an engine, it takes a pretty close second.

If you decide to proceed with rebuilding your own differential, you can use the information in this chapter along with some general and specific information in books like *High-Performance Differentials, Axles & Drivelines* by Joseph Palazzolo from CarTech.

Inspection

Inspect every part carefully when you have the rear differential apart. There are plenty of parts in the rear axle assembly that typically show signs of wear such as axles, bearings, clutches, and gear faces. There are also non-wearing parts that can have cracks.

DIFFERENTIALS

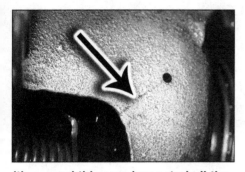

It's a good thing we inspected all the parts before the rebuild. We could have missed this hairline crack in the carrier, and it could have led to a catastrophic failure. This crack is not typical, but it drives home the point that inspection of non-wearing parts is critical. The hole was drilled to stop the crack from traveling farther so the carrier could be put back in service until it could be replaced because this has been a daily driver. (Photo courtesy Mary Pozzi)

There are different-size carriers for different-size ring gears. For the 10-bolt: 3.08 and numerically higher gear sets use a 3.08 carrier; gear sets that are 2.56 and 2.73 use a separate smaller carrier. For the 12-bolt: 2.73 and 2.56 gears use a small carrier; 3.08 through 3.73 gears used the medium carrier; and the 4.10 and numerically higher gear sets used the large carrier. The higher (numerically) gear ratio has a smaller diameter pinion head, so the ring gear has to be closer to the centerline of the pinion. That's why the carrier gear flanges are in different positions and thicknesses (see photo). Carriers from left to right, the 2-series, 3-series, and 4-series. (Photo courtesy Mary Pozzi)

Aftermarket Gear Sets

Strange Engineering and Precision Gear make special gear sets for 12-bolt differentials that are thicker—you can put a gear set that is 3.90:1 or numerically higher on a carrier originally designed for only 3.08:1 through 3.73:1 gears.

The 12-bolt rear axle is larger and stronger in every aspect when compared to a 10-bolt rear axle (below).

Unlike Ford 9-inch and older GM rear-differential assemblies, setting all the necessary clearances between the pinion and the ring gear of the differential carrier has to be done inside the rear axle housing. The Ford 9-inch and older GM carriers are a separate unit that can be unbolted from the rear housing and dropped out the front as one assembly. The Ford 9-inch assembly is comprised of the pinion, ring gear, and carrier, which allows all rebuilding to be done on a bench. Because the 10- and 12-bolt pinion is mounted to the entire housing and the carrier bolts in from the rear, the lash and pinion depth have to be set inside the housing. You can remove the housing from the car to assemble it, but that's not practical in most situations.

Gear Swap and Rebuild

The following is a detailed overview of a gear swap and rebuild of a Positraction rear differential. Depending on the specific repair or replacement being performed, you won't have to follow every step. Today, most enthusiasts do not rebuild the Positraction unit. Instead, many simply upgrade to another carrier, but we cover the rebuild, just in case.

The differential does not have a drain plug, so you must remove all the cover bolts to drain the fluid, but before you do, get a drain pan that will hold at least 5 pints. Often, the fluid smells very foul and will smell especially bad if the clutches have burned up in the Positraction unit. Remove all but two bolts, and leave them loose. A gasket scraper helps to pry the cover loose so the fluid can drain out. (Photo courtesy Mary Pozzi)

Differential Specifications

1967–1969	10-bolt cover	10-ring gear bolts	8.2-inch ring gear diameter	1.438-inch pinion shaft diameter	25-spline axles
1967–1969	12-bolt cover	12-ring gear bolts	8.875-inch ring gear diameter	1.625-inch pinion shaft diameter	30-spline axles

CHAPTER 7

Differential Rebuild

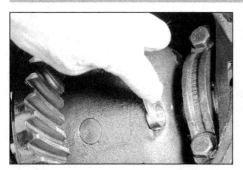

1 Spin the carrier around and find the bolt that holds the pinion shaft in place. Remove the bolt with a 1/2-inch wrench and slide the pinion shaft pin out of the carrier. Push the axles in, toward the center of the differential, about 3/8 inch. Then use a long needle-nose pliers or a magnetic-tipped probe to remove the C-clips from the ends of the axles in the center of the carrier. We found using a magnetic tool works great as long as there are no burrs on the tip of the axle. Pull the axles out carefully if you plan on reusing the axle seals. We strongly suggest replacing both seals since the seal lip typically gets damaged by drawing the axle across—having to pull the axle out again in a few weeks of service is a huge pain in the neck (or some other part of the human body). (Photo courtesy Mary Pozzi)

2 Before removing the bearing caps, mark one "right" and the other "left," or use a center punch to mark the caps with one dot for the left cap and two dots for the right cap (shown by the arrows), so that you don't mix them up during reinstallation. If you feel like you may still mix them up, dimple the inner gasket-mounting surface next to the cap with corresponding marks. Loosen the bolts about three turns and tap the caps with a dead-blow or other soft-faced hammer to get them loose, but do not pry them. They are specially machined and sensitive to stretching, much like an engine main cap. The yellow arrow shows the factory paint, which also helps show which cap goes on which side, if it has not been removed. The blue arrow shows the curved side of the cap which faces toward the outside case. The other side cap is square and faces the carrier.

3 If you're working on a differential in a car, loosely install the caps with bolts that are approximately 1.5 inches longer than stock. With the caps loosely installed, pry the carrier loose with a prybar (use a rag to protect the gasket surface). Be careful; you want it to pop out of the housing, but you don't want it to hit the floor and chip off a gear tooth. Don't pry directly against the gasket-sealing surface of the housing as seen here (maybe put a rag between the case and the prybar). Only pry on the outer case of the carrier. If there's a lot of preload, the carrier will not come free until it's almost completely out of the housing. Remove the caps and protect the carrier as needed. Have another person hold the carrier if possible.

4 If you're replacing the ring and pinion, remove the ring-gear now. Remove the ring gear bolts and tap the ring gear off the carrier with a soft-faced hammer if you plan on preserving the gear. Tap it off evenly around the perimeter; otherwise it will get damaged or at least get caught up at an angle.

5 If excessive wear or cracks are discovered, the carrier must be replaced; it's not worth rebuilding. Use a gear puller and washer over the end of the axle port to pull the bearings off the carrier. Make sure you're pulling against the bearing race as far inward as possible. There are two divots in the case for the feet of the puller to fit into. Lube the threaded puller shaft and tip on the bushing with anti-seize to ease use and protect the tool surfaces.

DIFFERENTIALS

6 Tap with a soft-faced hammer (or pry) the spring pack and plates out of the carrier with a brass drift until the two plates are out far enough to get a C-clamp on the two plates or run a couple of bolts through the springs to hold the pack together. The assembly only comes out through the large window. Put a second C-clamp on the center of the two plates or use a large set of Channel-Lock pliers to compress the assembly and remove it from the carrier with a screwdriver.

7 You can clamp an old, damaged axle into your bench vise to hold the carrier in place while you work. Turn the gears as needed by simply turning the carrier. Before removing more parts, mark all your parts and keep track of them. Be sure to mark the parts and shims that go together, so reassembly can go smoothly. Inspect all the bolts, shims, and components for excessive wear and damage. If the parts show any damage, they must be replaced. We had to clamp the carrier in a vise and drive the spider gear around with a brass drift through the window to get them to turn to the window so they could be removed. There should be a cup-shaped brass shim behind each spider gear.

8 Push the side gears and clutch packs out through the windows in the ends of the carrier. Use a hammer and brass drift if necessary. Remove both sets. Keep track of shims and clutch packs with the side gears they are associated with as they come apart. Pull the clips off the ends of the clutch packs with pliers and inspect the clutches for excessive wear.

New clutches are .070-inch thick. If yours are worn, replace them. You can add a shim between the carrier and the clutch pack to reduce clearance. Reassemble the clutches and plates on the side gear, starting and ending with a plate while alternating clutches and plates. Install clutch clips to hold the clutch plates together. Clips hold the clutches and plates together. These factory-style clutches have outer tangs that locate to the carrier by way of the clips, and the corresponding clutch discs have teeth in the center that lock to the side gears.

Differential Rebuild CONTINUED

9 Install one side gear, clutch pack, and shims. Then install both spider gears with their curved crush washers and the pinion shaft pin to keep the spider gears in place. Set a dial indicator on the face of the spider-gear tooth. Apply pressure to the side gear with a screwdriver and rotate the spider gear. The factory overhaul manual states that the tooth clearance should be between .001 inch and .006 inch. Disassemble and install shims in order to gain the desired clearance, then take it all apart and install the other side gear. Test the spider gear clearance for that side.

10 Install the clutch packs and both side gears along with the shims you chose for spider-gear clearance. Hold them in place with one hand while feeding the first spider gear into the carrier and spinning it around past the pinion pinhole. Now the tough part: Feed the second spider gear between the side gears and spin it around to the pinion shaft hole. We carefully rotated the side gear with a screwdriver. Use the pinion shaft to make sure the spiders are lined up with the hole in the carrier. Use a long screwdriver to assist during this process.

11 Factory spider gear thrust washers don't have a lip around the inner diameter. If you have replacement ones that do, you'll have to go back a step and install the spiders with their washers, which isn't easy. You can get them without the lip or carefully grind off the lips in a pinch to make your life much simpler. In order to get the spider-gear thrust washers between the spider gear and the case you have to spread the gears a little. A good tip is to use a couple of 1-inch-long fine-thread bolts (coarse versions are shown) with nuts as a homemade tool to spread them apart. The thrust washers are fairly soft, so be careful when installing them. Add a little gear oil on them before sliding them in. When both are in, confirm the pinion shaft pin fits through the assembly.

12 Compress the springs in the retainer plates with a couple of C-clamps or with large Channel Locks and feed it into the carrier between the side gears. If you're having trouble keeping the assembly together, run two 1/4-inch bolts through the plates and springs on the wide end until the assembly is installed far enough. The bolts must be installed in the direction shown or the gear flange will keep you from removing them as you slide the assembly into the carrier. Remove the helper bolts and drive the spring pack in. Then install the pinion shaft pin temporarily with the locking bolt threaded in to hold the assembly together until you're done installing the pinion gear in the differential housing.

13 Set the carrier in a press and drive the first bearing onto it with a heavy-duty sleave as shown. Then flip the carrier over and place a shim between the bench and the carrier case. Do not rest the assembly on the bearing cage, or you'll damage the bearing while driving the second bearing onto the carrier.

DIFFERENTIALS

14 It's a good idea to replace the ring-gear bolts with a new set of ARP bolts at this time. The original bolts have probably been over-stressed and should be replaced; but don't skimp on new bolts. Place the ring gear around the carrier. Before installing it, hand-thread two long bolts through the carrier into the ring gear. These help align the holes in the carrier with the threads in the gear during installation. Draw the ring gear onto the carrier with the actual ring-gear bolts. Tighten them evenly (with the exception of the long guide-bolts) in a star pattern to keep the ring gear parallel to its mounting flange during installation. Remove the guide bolts and install the correct gear bolts. Lubricated with 30W oil, torque the stock bolts to 50 ft-lbs.

15 While keeping the U-joint flange stationary, remove the front pinion nut. These nuts are for one-time use, so get new ones for reassembly. Using a brass drift and hammer, tap the front of the pinion into the differential. The U-joint flange comes loose during this process. Make sure the pinion doesn't damage the pinion bearings or its teeth during removal. Remove the front pinion seal and drive out the bearing spacer/crush sleeve. Both are one-time use only. (Photo courtesy Mary Pozzi)

16 Inspect all the bearings and races for excessive wear and replace as necessary. Inspect for wear on the pinion teeth and splines. If there are pits or rough faces on the teeth or burrs on the splines, you should be replacing the ring-and-pinion set. Also check the splines on the inside of the U-joint flange. Replace as necessary. We removed the races from the housing because we're installing a new pinion and all new bearings. Never install a new bearing without replacing its race also. (Photo courtesy Mary Pozzi)

17 Drive new bearing races into the housing for inner and outer pinion bearings. Use a bearing installer to safely install them straight and without damaging the bearing surfaces. Some race drivers are not long enough to install the inner pinion bearing. If this is the case, you may need to be industrious and make an extension in order to finish the job. (Photo courtesy Mary Pozzi)

18 Check the distance between the carrier bearing surface and the face of the inner pinion bearing in its race. Subtract the thickness of the pinion gear from the measurement and you have the shim thickness needed to put between the pinion and the inner bearing. Even after installing the shim and fully assembling the differential, it may be necessary to install a different shim if you can't get the correct tooth contact pattern on the ring gear. (Photo courtesy Mary Pozzi)

CHAPTER 7

Differential Rebuild CONTINUED

19 After installing the shim over the pinion, drive the inner pinion roller bearing using a pipe with an inside diameter slightly larger and 6 inches longer than the pinion shaft. Do this using a pipe large enough in diameter to only rest on the inner bearing race, but not so small that it touches the shaft of the pinion. Hammer on the pipe or put it in a hydraulic press. Do not forget to install the shim before you install the bearing. Use a special tool to remove the bearing from the pinion if a different-thickness shim is necessary. (Photo courtesy Mary Pozzi)

20 Slide the pinion bearing spacer (crush sleeve) onto the pinion shaft with the larger-diameter end of the sleeve toward the inner bearing. Slide the pinion assembly into the differential and slide the outer tapered roller bearing over the end of the pinion through the front of the housing. With a soft-faced hammer, drive the front U-joint flange seal into the front of the housing. Drive it in straight and only to the very edge of the housing lip.

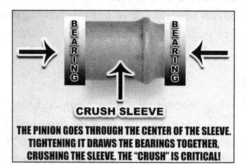

The crush sleeve is located on the pinion gear shaft between the inner and outer bearings. The amount of crush is what sets the clearance on these two bearings. If it's not crushed enough, the bearings will have too much clearance and vice versa if the sleeve is too crushed.

21 Put Permatex Ultra Black sealer on the splines inside the U-joint flange to help keep fluid from seeping out. Put a little oil on the outside sealing surface to lubricate the outer seal. Use an inch-pound torque wrench. Install the new washer and nut on the pinion shaft and tighten to 20 to 25 inch-pounds (not foot pounds) for new bearings and crush sleeve and 5 to 10 inch-pounds for used parts.

22 To space the carrier in the housing, GM used a single, thick, cast-iron spacer ring (on left), but new carrier bearings are slightly wider, so the ring is not necessary. Lubricate the carrier bearings with gear oil and install the carrier with its races with bearing cap on the right side, hand tight. Be careful not to damage gears during the process. Measure the thickness of the shims needed to install between the left bearing and the housing, and slide them in. Hand-tighten the left bearing cap and remove the right bearing cap. Measure for the thickness of the shims necessary to install between the right bearing and the housing. There will be an additional .008 inch of preloaded to the shims installed. Slide the shims into place. The last one needs to be carefully tapped into place with a soft-faced hammer. Be careful not to fold or bend the shim during installation. When installing new shims, slide the two thicker ones in all the way and then tap the thinner, correct-thickness shims between them to ease installation. Shim installation forks are available to help cup the shim while driving it in, but you can make your own with some steel plate. Install the bearing cap and torque it in place.

DIFFERENTIALS

Backlash

To get the best performance at traction from the differential, you need to correctly set the backlash.

Check the ring-gear-to-pinion backlash with a dial indicator. The factory manual suggests a range from .003 inch to .010 inch, but the preferred lash is .005 inch to .008 inch. If backlash is not within specification, move the shims from one side of the carrier to the other side. Do not take the shims out; there must still be .008-inch bearing preload. If you move a .002-inch shim from left to right (in order to reduce backlash), you reduce backlash by .001 inch.

Reposition the dial indicator to read the runout on the ring gear. It should not exceed .002 inch, but if it does, check for foreign debris in the bearing caps or cap deformation.

Mark the ring gear with gear-marking compound on four teeth in two separate locations. Turn the ring gear and inspect the pattern left by the pinion. The pattern changes shape and location according to the shims you've installed between the carrier and the differential case, as well as the shim you placed between the pinion and the inner pinion bearing. To get the correct pattern, change shims in those locations.

Install new axle bearings and seals in the ends of the differential housing.

Install brake brackets or backing plates if you removed them.

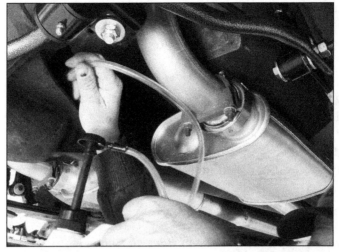

Don't forget one last and very important detail. Fill the differential with fluid. According to the Chevrolet Chassis Overhaul Manual, the rear axle has lubrication capacities of: 10-bolt, 4.5 pints; 12-bolt, 4.9 pints. But the best rule on filling the fluid is to have your Camaro as close to level as possible, unscrew the fill plug, fill with the correct fluid for your application, and stop filling when the fluid starts draining out of the fill plug hole. Don't forget to reinstall the fill plug and tighten it to 20 ft-lbs. (Photo courtesy Mary Pozzi)

Remove the spider-gear alignment pin and slide the axles into the housing one at a time. Slide them in far enough to install the C-clip over the tip of the axle and pull the axle back out to lock the C-clip in place. Do the same for the other axle. Install the spider-gear pin and the pin-locking bolt and torque it to 20 ft-lbs.

Confirm there aren't any contaminants in the housing. Clean the gasket surface with some brake cleaner.

Install the differential cover with a new gasket with a little Permatex gasket sealer.

Test Drive

After rebuilding a differential, take the car out on the city streets and drive it for a good 30 minutes to break in the gears. No burnouts or high-speed runs on the freeway; just cruise around town getting up to 35 or 40 mph, coasting and decelerating. This allows the fluid to get between the teeth and helps break in

Drive the axle bearings and seals into the ends of the housing with a seal and bearing driver. The driver pictured is homemade, but you can get drivers at most tool stores. Put a light film of gear lube on the lips of the seals. Being careful not to damage the seals, slide the axles in and install the C-clips and carrier pinion shaft. Install the rear cover and install all your brake components. (Photo courtesy Mary Pozzi)

CHAPTER 7

the gears without damaging the surfaces of the teeth.

Identifying Parts

As with the engine and most mechanical parts of the Camaro, the factory had a system of cast-in and stamped letters and numbers to identify what was inside the differential. If your housing has all the original internal parts, you can figure out exactly what's inside without popping the rear cover off. Because changing gears is not a simple task, the odds are in your favor that the internals will match the stamps. The external casting numbers only identify the casting date and type of differential housing, which is only a small piece of the puzzle.

Casting Identification

There are numbers cast in the housing on the front and back sides of the cast-iron center section—casting date, foundry information, and testing numbers 1 through 10 (listed as 0). The castings changed over the three production years, so some of the casting numbers didn't exist on every housing, and some of their locations even changed.

Date

Casting date info on the upper left side of the housing (when looking at the differential cover) didn't change location. The breakdown is as follows:

- The casting month is designated by the letter (A = January, B = February, etc.).
- The next two digits represent the day of the month.
- The last digit is the year (7 = 1967, 8 = 1968, 9 = 1969). Note that because production of parts started in 1966 for the 1967 model, many 1967 differentials have a 6.

Foundry

The foundry cast and stamped ID numbers into the block to clearly identify both small- and big-block engines. You need to locate and decode these ID numbers, so you can verify the particular engine's size, date of manufacture, and equipment package.

Test and Housing Numbers

The casting test numbers were not on all differentials. We've seen 1967 housings without them, but 1968 and 1969 housings have them to the left of the housing cover.

On the lower-left support rib on the front side, there are numbers cast into the housing to designate a 10- or 12-bolt differential:

- 3894860 = 10-bolt
- 3894859 = 12-bolt

You can identify the differential without this number. Simply count the number of bolts holding the cover on the differential.

The upper-left-side digits provide the casting date of the differential center section. The letter K is for the month of November, the 27 is the day of the month, and the 8 is the last digit of the year it was cast. This housing is in a 1968 Camaro, but the late 1968 cast date means that it came out of a 1969.

Housing Stamps

The factory housings all have letters and numbers stamped on the front of the right axle tube. These numbers are known as "axle codes," and designate the production date as well as the factory gear and carrier options in the differential. For the largest and most accurate information for differential numbers, visit www.chevroletcamaros.com Its databases are constantly updated with current casting and stamping data as new information becomes available.

Identifying Gear Ratio

It's always possible the gears on your car are not stock and the stamps on the housing tube won't match. Therefore, the only sure-fire way to identify the gear ratio is to count the teeth on the ring gear and the pinion gear. Then divide the number of ring gear teeth by the number of pinion teeth. For example:

If you have 41 ring-gear teeth and 15 pinion teeth, 41 ÷ 15 = a gear ratio of 2.73:1.

Original GM gears don't have gear ratios stamped in them, but some aftermarket gears do.

This 1969 axle code breaks down as QX = 3.73 posi, 05 = May, 05 = assembled on the 5th day of the month, G1 = Detroit Gear and Axle 1st assembly shift. The second line has the letter "E" for Eaton as the Positraction manufacturer. (Photo courtesy Bob Jones)

CHAPTER 8

BRAKES

You need to determine the restoration grade (concours, daily driver, survivor, etc.) and level of authenticity for your Camaro. With a complete concours restoration, retain the stock brakes, including rear drum brakes. If you've found a very rare car with a JL8 rear disc brake setup, retain these brakes also. However, for most owners, particularly of common 6-cylinder and V-8 Camaros, an aftermarket brake upgrade delivers a tremendous performance improvement and doesn't degrade the value of the car. Since there are so many options and brakes can easily switched back to a stock setup, we also cover some practical applications and aftermarket upgrades for street cars and daily drivers.

It's rare to see a set of original JL8 rear brakes since only 206 Camaros came with this option from the factory and it was a one-year-only RPO. The JL8 package was produced in an effort to win Trans-Am races, even though Chevy wasn't "racing." (Photo courtesy Bob Jones)

Original Components and Restoration

The original master cylinder, brake booster, residual pressure valve, and distribution/safety switch block in our 1968 SS/RS project car were original, but needed some serious work to get them back to good operating condition and being esthetically pleasing. Power Brake Exchange (PBE) in San Jose, California, was up to the restoration task. Sure, you can purchase cheaper replacement parts, but not only are they not original to the car (if you care), but the replacement units are not usually visibly the same. Having the original parts rebuilt ensures they will bolt back on the car correctly, and their appearance won't differ from original.

PBE has been restoring brake components for a long time and they really knows its stuff. All the parts were painted or plated to their original factory finishes. The booster was disassembled, replated in original

CHAPTER 8

For an original look, the master cylinder is flat black with the machined boss faces bare machined finish. Your options are to paint the machined surface with cast-iron-colored paint or add a light coat of gun oil to keep the surface from rusting. The bail wires (straps) on the master cylinder are bare steel or silver if you want to paint them and the top cover is gold-cadium plated.

This 1967 manual brake master cylinder is original and is supposed to be finished the same way as the power-assist unit. Manual masters don't have the hold-off valve on the side and the distribution block underneath is plumbed differently.

gold cadium, and rebuilt with all the original steel parts, which were confirmed by checking the factory numbers stamped on the locking tab. The master cylinder was rebuilt with new internal parts after they bored out the cylinder and sleeved it with a stainless-steel sleeve. Afterward they restored the outside with factory-style finishes on the clamps and cap. The safety-switch block was blasted and the brass flares were replaced. The residual pressure valve went through similar processes.

Brakes are a vital safety system and you must ensure your brakes are in proper working order. If the Camaro's pads or drum liners are worn past minimum thickness or the rotors are warped, perform a brake service job. Because many brake pads are made of asbestos, be sure to wear gloves and a respirator when servicing the brakes. (We won't cover each step in brake disassembly, inspection, and assembly, but you can find the information in a factory service manual or *High-Performance Brake Systems* by James Walker.)

Brake Hardlines

There are many different configurations of brake hardlines because so many different manual- and power-brake systems were installed in first-generation Camaros. The original brake lines are made of a mild steel. You can buy new brake lines from many reproduction companies in the original mild steel or you can buy the same lines in stainless steel for a little more money. The stainless steel looks great and lasts forever, but it won't look exactly like the factory finish.

Long brake lines come bent in the middle and stuffed in a huge box, so there will be a little modifying to get them straight, but with an inexpensive bender you can get them straight fairly easy. To get these full-length brake lines in the car, you're much better off doing it with the body off the frame, even only a few inches. Otherwise, you'll scratch the heck out of your car and probably damage the new brake lines. A little bending is usually necessary, but with the good kits (from Fine Lines) they take original lines out of unrestored cars and precisely replicate them. With this kit every piece lines up exactly the way it is supposed to, unlike lines we've used from other well-known companies.

Master Cylinder

It's likely that your master cylinder has been replaced and is not original. (If it is original, you can have it restored with a new sleeve and seals.) Master cylinders have a different-size bore for each application, and there are at least three different sizes: 7/8, 1, and 1 1/8 inches.

Each original master cylinder has a two-letter stamp on the small machined pad in the front. There have been at least seven different stampings documented on master cylinders. Those stampings are: AD, AU, BS, CT, SA, US, and WT. Many of the master cylinders were physically different and varied by application, but some were only different internally by the bore size required for the given application. The only way to differentiate these similar-appearing masters is by the two-letter stamp. The master cylinders ranged from: 7/8-inch (AU-manual J65 in 1967) to 1-inch (BS-manual and power drums for 1968, and 1969, AD-manual front disc in 1967, and CT–manual and power drums with mixed usage) to 1 1/8 inches (WT-power front disc 1967 and 1968 and US-power front disc and power 4-wheel disc [JL8]).

The master cylinders were originally painted flat black with the machined surfaces (front stamped pad, bosses around fittings, and the tapered hole in front [on some]) left bare cast iron. The bail wires (master cylinder cover retainers) were left bare steel (silver in color). The covers were gold-cadmium plated. Bare cast iron rusts in a very short time, so put on a light coat of gun oil or paint the surface with gray cast-coat.

Hold-Off Valve

According to the Chevrolet *Chassis Service Manual,* the valve assembly located between the master cylinder and the brake distribution block on the front brake circuit is called a "pressure-regulator valve." With that said, the more operationally correct name for this valve is "hold-off valve" and for this book I refer to it as the hold-off valve because the valve on the side of the subframe under the car is more appropriately called a pressure-regulator valve.

To operate, disc brakes require more than three times the amount of line pressure compared to drum brakes. Chevrolet added the hold-off valve to manual and power-assisted front disc/rear drum–equipped cars, so the front brakes wouldn't come on too strong (compared to the rear drum brakes) under light braking. The valve keeps the pressure going to the front disc brakes until approximately 35 pounds of pressure is applied to the rear drums. A front-disc-equipped Camaro nose-dives excessively if the valve is missing or stuck open. Some self-appointed experts claim some disc/drum cars did not have a hold-off valve and that some 4-wheel drum cars had them, but it's difficult to know everything the factory did, and new information surfaces all the time.

The factory finish on the hold-off valve (under the master) is semi-gloss black with a cast-iron finish on the large nut and the circumference of the round boss on the regulator body. The small bracket attaching the valve to the master cylinder is gold-cadmium plated and the bolt attaching the bracket to the valve is zinc plated.

Brake Distribution Block (with brake warning)

The first-generation Camaro factory brake distribution block has front and rear brake lines going into and out of it, as well as a single wire for the brake warning switch. The distribution block is located about 1 inch under the master cylinder and mounted to an L-shaped bracket that's attached to the right-front master-cylinder mounting stud. The distribution block and its bracket were painted semi-gloss black from the factory.

There are at least two different distribution blocks: one that has a cast front face and one that has a machined front face (machined surfaces did not have paint). The latter was used on first-generation 4-wheel drum, front disc/rear drum, and 4-wheel disc brake applications. This block distributes the single front brake line that comes from the front-brake circuit of the master cylinder. A pressure-regulator valve on disc-brake-equipped cars is attached to the block and splits the fluid into two lines, one for the left front and one for the right front brake assembly.

The rear brake line is a single line coming from the rear circuit of the master cylinder. It comes back out of the distribution block as a single brake line to the rear brakes.

The brake warning switch is located in the middle of the distribution block.

The front-circuit fluid going into the block pushes on the piston in the center of the distribution block; the rear circuit fluid pushes on the piston from the other side. The equal pressure on the switch actuator piston from both sides helps center the piston. If the rear circuit fails (i.e., rear brakes lose pressure from a ruptured brake line), the pressure on both sides of the piston is no longer equal, causing the greater pressure on the front brake circuit to push the switch actuator piston inside the distribution block to an off-center position (toward the rear in this example).

Once the piston is off center, it actuates the brake warning switch, which turns on the little red warning light in the dash cluster. In some cases, if the piston travels off center far enough, it can close off the failed half of the circuit (rear brakes in this example) so the good half of the system (front circuit in this example) will still stop the vehicle. When one circuit is closed off, the brake pedal still feels firm, possibly more firm than if both circuits are working properly, so if the brake warning light is on, you better find out why.

Once you fix the leak in the system and bleed the brakes, the pressure in the distribution block equalizes. As a result, the switch-activating piston is pushed back into the center and the brake warning light turns off.

Late-model cars have a similar-looking block under the master cylinder. The new blocks not only contain the brake warning switch, but also contain proportioning valves (and residual-pressure valves on rear drum brake applications) and are called combination valves. Be aware that not all parts from newer cars or other cars from the same year work in conjunction with the parts on your car and application because of the existence of valves inside of components that you cannot see.

Pressure Regulator Valve

The pressure regulator valve is located under the left front seat on the outside of the subframe. The valve limits pressure going to the rear brakes, which is regulated by the pressure on the inlet of the valve. This helps the rear brakes from being overdriven during hard braking, which could cause a rear brake lock-up condition. The factory finish on the valve is machined brass. It's a Kelsey-Hayes valve with the letters "KH" stamped on it, along with an "R" and a date code.

This valve was infrequently installed on the Camaros. According to the instructions in the 1967 factory *Camaro Assembly Manual*, the pressure regulator valve was installed on Camaros with air conditioning (RPO C60). The manuals for

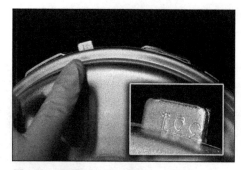

The front of the top tab on the original brake booster is stamped with the day of the year on which it was produced. An original booster-day number should closely match the build date on the cowl tag. The 331st day is November 26.

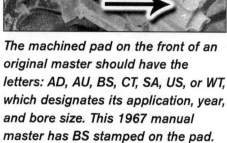

The machined pad on the front of an original master should have the letters: AD, AU, BS, CT, SA, US, or WT, which designates its application, year, and bore size. This 1967 manual master has BS stamped on the pad.

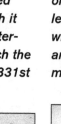

To get down to the fine details, Super Car Workshop restored these original wheel cylinders to their original look with paint to match the casting color and replaced the green stripe, which was barely visible when the car was disassembled for restoration. (Photo courtesy Brian Henderson)

The old brake booster was sent off to be rebuilt. While it was disassembled the shop had the parts gold-cadmium plated for a factory-looking finish.

The 1967 Camaro was available with the RPO J52 (4-piston caliper, 1 inch thick, and 11-inch-diameter rotor). The 1969 J52 was changed to a single-piston caliper. it is shown here with correct factory finishes.

The JL8 front brakes were completely different than the J52 units. The name "ArmaSteel" means it was cast by the same company that cast crankshafts and other components. The front rotors are 1¼ inch thick and have an 11¾ inch diameter. (Photo courtesy Bob Jones)

BRAKES

Always take a picture of the drum-brake springs, clips, and other components, before removing any of the parts. If you're good at puzzles, you may not have any trouble rebuilding the brakes, but it's not usually worth taking a chance.

1968 and 1969 say the valve was installed on SS models (RPOs L48 and L78). But both of the 1967 and 1968 *Chevrolet Chassis Service Manuals* state the valve was used on RPO C60 Camaros. And in the 1969 service manual the valve is listed for use on Camaro models equipped with 8-7/8-inch ring gears (12-bolt differentials).

We've seen the valve on 1967 427 Yenkos, which are not C60 equipped but have a 12-bolt rear axle. We've also seen them on non-SS models like the 1969 LM1, which was the first non-SS Camaro to get a 350 engine; so it's possible that other performance models like the 1967 and 1968 L30/M20 (275-hp 327 models matched with M20 4-speed transmission) also came with this pressure regulator valve and that the factory set a standard or rule for which Camaros received the valve, but it didn't make the corrections to or notes in the assembly or service manuals.

OEM Brakes

Compared to today's standards, the original 1967 through 1969 brake systems were substantially insufficient with manual front and rear drums to adequate brake performance with the rare power disc front and rear. It wouldn't hurt the value of most cars to at least upgrade to factory front discs and rear drum brakes, and it's an important safety upgrade.

4-Wheel Manual Drum (Standard Brakes)

All first-generation Camaros came standard with four-wheel manual drum brakes. For the time and low-performing lower model Camaros, the standard brakes were adequate, but left a lot to be desired in performance applications. The front drum brakes had a small hub with a separate brake drum that slid over it.

J50 4-Wheel Power Drum Brakes

The lowest level of brake upgrade is the RPO J50 where the four-wheel drums were vacuum-assisted power brakes. This RPO was offered all three years of the first-generation Camaro.

J52 Power and Manual Front Disc Rear Drum (1967 through 1969)

The most common power front disc and rear drum is the RPO J52. The J52 brakes changed from a 4-piston fixed-mounted front caliper in 1967 and 1968 to a floating single-piston caliper in 1969. The 1967 and 1968 rotor was a two-piece rotor and switched to a one-piece rotor in 1969. The rotor was 1 inch thick and 11 inches in diameter.

J56 Power Front Disc Rear Drum (1967 Low Production)

Corvette 4-piston brake calipers were adapted to the work on the front of the Camaro during the process of Chevrolet's development of racing brakes. These low-production (less than 500 units) front brakes were available for 1967, which required Z28 RPO. The rotors were 1¼ inches thick and 11¾ inches in diameter. The calipers are easily identified as J56 because of the two pins holding the pads in place, whereas the JL8 caliper uses a single pad-locating pin. This system utilized HD metallic rear brake shoes to increase rear braking. The J56 production numbers are even lower than the rare JL8 4-wheel disc brake package.

J65 4-Wheel Drum Brake Metallic Pads

This was a one-year-only (1967) option. The RPO J65 was a drum brake shoe upgrade to HD metallic linings for Camaros equipped with 4-wheel drum brakes. It was typically reserved for performance-level SS-equipped Camaros.

JL8 4-Wheel Power Disc Brakes (1969 Low Production)

One of the rarest of the rare options is the JL8 four-wheel power disc brakes. According to historians, there were only about 206 Camaros that left the factory with the JL8 brake option when it was offered as a regular production option for a short period of time for the 1969 model year. Before then, during the 1968 model year, the JL8 was only available at the dealership as heavy-duty service equipment. All of these facts add up to the JL8 being very rare and if you have a JL8-equipped car (by the factory or service upgrade), you've got yourself some really awesome and valuable equipment.

If you don't have the stock JL8 4-wheel disc brake equipment, but want to make the investment to install them on your Camaro you can find the parts at www.jl8brakes.com.

Home-Brewed 4-Wheel Disc Brakes

Maybe you want to build your own 4-wheel disc brakes from non-first-generation Camaros or even non-GM brakes to save a little money. We've done it with less-than-stellar results. You are taking your life in your own hands if you try to design a system with mismatched brake parts. There are a few places on the Internet that explain how to upgrade your brakes by using late-model parts, but for safety reasons we can't suggest using their information. Maybe, these guys knew what they were doing, or they got really lucky. There is a lot of engineering, mathematics, and fluid dynamics knowledge that goes into building a well-balanced brake system.

It's important to have a balanced brake system, which means that the front brakes don't overpower the rear brakes (too much) and the rear brakes don't overpower the front brakes. Through pressure from depressing the brake pedal, the master cylinder moves fluid to each wheel. If 2 ounces of fluid move the front caliper pistons 4 mm and the same pedal position pushes 2 ounces of fluid to the rear calipers (which for this example have larger bores than the front brakes), it moves the rear caliper pistons .5 mm.

A system like this never gives you enough rear caliper movement to be effective. In fact the front brakes are doing all the work and the rear brakes will probably never work. This scenario could not be safely remedied by installing an adjustable brake-proportioning valve, because you'd have to limit the amount of brake fluid going to the front brakes, which would be a ridiculous fix. If you're going to design your own brake system, educate yourself.

Aftermarket Components

Stock drum brakes leave a lot to be desired in performance. If you've never driven a 4-wheel drum brake car, or haven't driven one in the past 10 years, you may not know or recall that they do not stop very fast and are really just not safe for any kind of performance driving. If you're looking for better braking than some old stock four-wheel drum brakes or want to upgrade from some old crusty stock disc brakes, there are many options on the market. If you have factory four-wheel disc brakes on your first-generation Camaro, please don't swap them for aftermarket brakes; if you do, contact us and we'll find somebody to take those JL8 brakes off your hands.

Because this book is for practical restorations we felt that it only makes sense to include information on all brake options, including aftermarket brake systems. I think it's safe to say that most of the people restoring/rebuilding their first-generation Camaro might not be able to afford to spend approximately $8,000 dollars to install JL8 front brakes and an additional $5,000 for JL8 rear brakes (or $15,000 for the JL8 rear brakes with correct differential assembly) to complete the only first-generation factory-offered 4-wheel disc brake package. You could spend $13,000 or $23,000 on factory-correct (or close to it) 4-wheel disc brakes, but many can't afford that. Most people opt for aftermarket high-performance disc-brake conversions that equal or exceed the performance of OEM-model JL8 brakes.

That's where aftermarket companies, such as Baer Brakes and CPP Brakes, come into the picture. They both offer kits that fit inside the stock 15-inch wheel and still have an integral parking brake in the rear calipers. For an example we chose Baer brakes.

Typically, when you think of Baer Brakes, you think of brake rotors so big that you have to run 17- or 18-inch-diameter wheels to clear them. Well, Baer Brakes also offers a smaller-diameter disc brake system to fit factory wheels and deliver high-performance braking. Its Serious Street package with one-piece rotors and Serious Street Plus with two-piece rotors are its answer for muscle car owners who don't want to upgrade to 16-inch-or-larger diameter wheels. The front rotors are 11 inches in diameter, the rears are 11.35 inches in diameter, and they fit under 15-inch factory-style steel wheels. Smaller 14-inch wheels are simply too small, so 15-inch wheels are required.

Baer was successful at mixing late-model Corvette brake-caliper design and performance with a special combination of rotors with smaller inner-hub diameters than its other Sport, Track, Extreme, and other performance brake systems. Some aftermarket brake companies offer single-piston disc brake front kits, which resemble stock 1969 Camaro front brakes and a later-model (non-early) Camaro-appearing caliper so the parking brake cable can be incorporated. The older single-piston caliper lacks the performance offered with the Baer kit, which has a better-performing two-piston front caliper along with the late-model caliper in the rear. It only makes sense to upgrade to the best brakes for the bucks if you're going to upgrade the brakes at all.

BRAKES

Disc Brake Installation

1 Put the car safely on jack stands. Remove the wheels. Place the new caliper over the new rotor and place them inside the wheels. This is to ensure that the new brake caliper and the center of the hub fit the wheels before going through the trouble of installing the whole system. Do this for front and rear systems. (Photo courtesy Mary Pozzi)

2 Loosen the brake fitting where it connects to the brake hose at the side of the subframe. Use the correct 3/8-inch line wrench, or you'll round the edges off the nut and turn this into a much larger task, which requires vise grips and replacing the nut and/or brake line. Spray the fittings with penetrating oil if necessary. (Photo courtesy Mary Pozzi)

3 Somebody had already installed this non-original master cylinder. We're replacing it with the one Baer included in the kit. The master cylinder bore size is important, so we popped the lid off and allowed the brake fluid to drain out of the front lines into a container, then put the lid back on for now. We installed the included caps on the end of the brakelines to keep out moisture and debris. (Photo courtesy Mary Pozzi)

4 Remove all of the brake components from the spindle, but don't remove the spindle. Our car has disc brakes, so we pulled the bolts out of the caliper, pulled the brakeline clip off the bracket at the frame, then removed the caliper. Remove the dust cap, cotter pin, and spindle castle nut. Pull the rotor off the spindle, and clean the grease off the spindle pin, so you don't get unnecessarily greasy. Remove the two bolts holding the steering arm to the spindle and the last large bolt holding the backing plate to the spindle. If you have drum brakes, pull the drum off the separate hub. Next, you must completely disassemble the brake shoes, components (which are easier to remove if you have the proper drum brake tools), and hub before you can remove the backing plate. (Photo courtesy Mary Pozzi)

CHAPTER 8

Disc Brake Installation CONTINUED

5 Install a new caliper bracket and original steering arm on the spindle with the supplied 1/2-inch hardware. The longer bolts are for the thicker portion (the rear) of the steering arm. Torque them to 90 ft-lbs. With a new rotor, put a film of assembly grease on the inner seal and then slide the rotor onto the spindle. Put the supplied outer wheel bearing and keyed spindle washer onto the spindle. Install the castle nut finger-tight while turning the rotor. If the threads on the spindle are bad, replace it. Tighten the nut with a wrench while rotating the rotor, but as soon as you feel the bearings stop spinning freely and you feel drag on the rotor, back the nut off until the drag stops. (Photo courtesy Mary Pozzi)

6 The cotter pin should fit through one of the recesses in the castle nut and through the spindle pin. If the nut has to be moved to get the cotter pin through the hole, do not tighten the nut so that the rotor drags on the bearings. Instead, spin the nut counter-clockwise (loosen) until the cotter pin slides through the castle nut into one of the holes in the spindle pin. When the pin is installed, bend over the ends of the pin. Installing the cap was made easy by using a short pipe that was the correct diameter to fit on the lip. As with any brake install, use brake cleaner to clean the brake surface of grease and oil before installing pads or shoes. (Photo courtesy Mary Pozzi)

7 Grab the correct caliper with its anchor plate for the correct side. The brake-hose hole is mounted low and the bleeder screw is mounted high (so the air bubbles rise to the top and easily bleed out). Slide the caliper over the rotor and attach the caliper and its anchor plate to the billet bracket with the supplied metric bolts to 110 ft-lbs. (Photo courtesy Mary Pozzi)

BRAKES

8 Use the supplied lock clip to secure the brake hose to the factory frame bracket. Attach the other end of the hose to the caliper so that it does not bind, twist, or kink during the full range of suspension travel and turning from right to left. (Photo courtesy Mary Pozzi)

9 Make sure you have a copper crush washer between the caliper and the banjo fitting and another crush washer between the banjo fitting and the banjo bolt. These are not regular washers, so don't use any other type of washer in their place. Torque the banjo-bolt to 20 ft-lbs. WARNING: Banjo bolts are designed to break if they're overtightened. Perform the same procedure on the opposite front wheel. (Photo courtesy Mary Pozzi)

10 Remove the brake drums. If they don't pop off the ends of the axles, they may be rusted/fused to the axle flange or the centering hub on the center of the axle. In this case, a sharp hit with a rubber or plastic mallet typically breaks it loose. If the drum is loose from the axle but won't come off, it's possible that the brake shoes are stuck in a groove worn in the brake drum. If this is the case insert a long flat screwdriver or a brake-adjustment tool through the slot in the face of the brake drum and adjust the rear brakes to get the brake shoes away from the brake drum. (Photo courtesy Mary Pozzi)

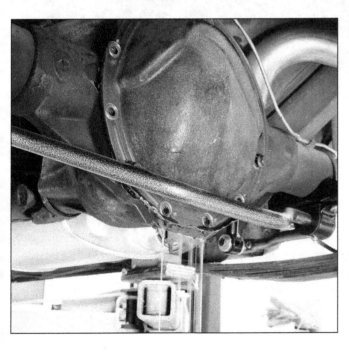

11 In order to remove the drum-brake backing plates from the rear axle you have to pull the rear axles out of the housing. Place a drain pan under the differential cover and remove the cover. Loosen and remove all the bolts except two. Drain the fluid by loosening the cover. (Photo courtesy Mary Pozzi)

12 With the cover off, remove the pin-locking bolt and then slide the locking pin out of the gear carrier. (Photo courtesy Mary Pozzi)

Disc Brake Installation CONTINUED

13 Push the axles in one at a time and slide the C-clip off the end of the axle. Carefully slide the axle out of the axle housing without tearing the axle seals. (Photo courtesy Mary Pozzi)

14 Inspect the axles on the outer bearing surface (arrow shows worn surface) to make sure they're in good shape and that they're not bent or twisted. Replace an axle if there's any sign of trauma. Now is also a good time to replace any broken wheel studs. Check the axle seal as well, but this is actually a good time to simply replace it because it could be the original seal. (Photo courtesy Mary Pozzi)

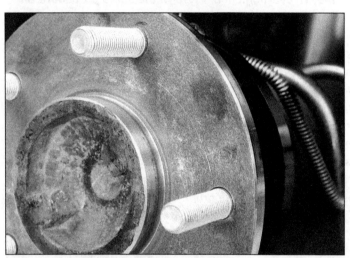

15 Make sure the outside diameter of the axle flange is 5.9 inches or less. Also make sure the center hub surface is the correct diameter. If it's larger, the rotor won't seat against the axle flange and will cause caliper-alignment issues. Take the axles to a machine shop to have the flange diameter reduced if necessary. Here you can see how the hub and outer diameter have been machined to reduce their diameters. (Photo courtesy Mary Pozzi)

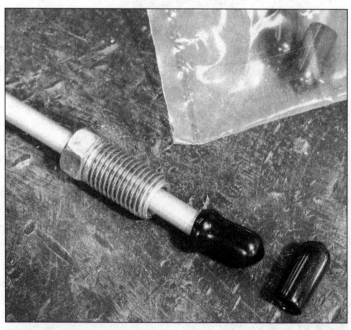

16 Using your line wrench, remove the hardline from the back of the wheel cylinders and cap the line ends with the supplied caps to keep debris and moisture out of the lines. (Photo courtesy Mary Pozzi)

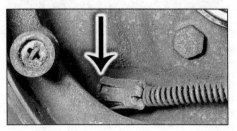

17 Remove all the brake shoes and accessories from the backing plates. Using correct drum brake tools makes the job much easier. Remove the parking brake cable by removing the adjustment guide from the inside of driver's-side frame rail. Pull the cable ends out of the junction brackets. Remove the rear parking brake cables from the backing plates by depressing the clip tabs. Remove the backing plates from the housing ends. (Photo courtesy Mary Pozzi)

BRAKES

18 Install the supplied rear axle brackets on the housing ends using the original backing-plate bolts. The bracket is mounted so the caliper bolt flange is pointed away from the shock absorber, otherwise the caliper hits the shock. Test mount the correct caliper (bleeder fitting pointed up) to the axle bracket. The supplied parking-brake cables should mount on the caliper assembly and route just underneath the leaf spring. If you have aftermarket equipment on the car, you may need to adjust the axle brackets accordingly. Torque the flange bolts to 35 ft-lbs. (Photo courtesy Mary Pozzi)

19 Put a light film of assembly grease on the axle seal and install the axles in the correct side. Slide them in and install the C-clips, then move them out to hold the C-clips in place. Install the pin back in the gear carrier and install the locking bolt, but don't torque it yet. (Photo courtesy Mary Pozzi)

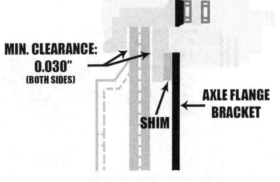

20 Determine the correct rotor direction. Install the rotors on the axle flanges and use two offsetting lug nuts to hold the rotor firmly against the axle flange, so it does not fall on the ground. (Photo courtesy Mary Pozzi)

21 Use the supplied metric bolts to hold the caliper to the axle bracket and snug them down for measuring clearances. Using the bracket-alignment diagram as a guide, check the caliper anchor bracket clearance to rotor clearance. Some variances exist in stock and aftermarket parts, so make sure there's a minimum of .030-inch clearance between the rotor and each side of the bracket. Shim the anchor bracket as necessary with the supplied shims. If the clearance is too small between the inside rotor face and bracket, the bracket may need some machining (contact the Baer tech department for details). Torque the metric caliper bolts to 85 ft-lbs when the clearances are correct.

22 Baer includes its own hardline retainer kit, which includes tabs to weld to the rear axle housing tube. You can use the instructions, or (if you don't want to make a change that would take some undoing if you wanted to go back to stock) use hose clamps around the axle tube to anchor the holding tabs. Attach the supplied brake hoses to the calipers with the same method of crush-washer placement as used on the front calipers. Attach the other end of the brake hoses to the original hardline. Make sure the routed hoses do not interfere with moving parts during suspension articulation, torque the banjo bolts to 15 to 20 ft-lbs, and cinch the brakelines to the hardlines. (Photo courtesy Mary Pozzi)

CHAPTER 8

Fully clean the spindle and check for damaged threads as well as possible damage on the surface where the bearing sits. We've seen inner wheel bearings lock up and seize to the spindle and damage the spindle pin. If you find any cuts or damage on the spindle pin, look for a replacement. Install the new caliper bracket and the original steering arm using the supplied bolts, and torque them to 90 ft-lbs.

Grab the correct rotor for the proper side of the car. The rotors have gas-release slots on the outside face and cooling vanes inside. Both features require specific directional rotation to work correctly.

Before working around brakes, especially drum brakes, you should be aware that there is a lot of loose brake dust that could get on your skin and in your lungs. Over the last 40 years, brake linings have been made from materials known or suspected to cause health problems. Drum brakes are the worst for brake lining dust because large quantities get trapped inside the drum. Striking the drum with a hammer to help remove it from the axle flange or simply removing the drum and turning it over releases a cloud of brake dust. I will be very clear: You need to wear protective eye wear, a proper respirator, and protective gloves when working around brakes until you've thoroughly wiped away all the brake dust.

Install the Master Cylinder

Make sure you have a bucket under the lines and have some disposable rags readily available because brake fluid eats through paint faster than you can wipe it off. Do your best to keep the fluid contained and keep your fender covered with plastic or

Disc Brake Installation CONTINUED

23 *Install the new Baer parking-brake cables to the calipers and the other ends to the factory emergency-brake cable brackets on the frame. Install the other parking-brake cable components and adjust the cable tension. (Photo courtesy Mary Pozzi)*

24 *Once all the clearances and parking-brake cable routing have been verified, torque the carrier pin-locking bolt to 20 ft-lbs. Use a new gasket with a thin film of Permatex Ultra Black sealer on the housing and cover mating surfaces, and torque the cover bolts to 20 ft-lbs. Unscrew the fill plug from the differential and fill it with the proper fluid for your Positraction or non-Posi rear axle. With the car level, fill until the fluid starts to come back out of the fill plug and put the fill plug back in and tighten it. (Photo courtesy Mary Pozzi)*

BRAKES

Now is a good time to have your booster rebuilt and re-plated with gold cadmium. You're much better off having your booster rebuilt if it's an original part. However, if you're removing it and don't want to keep it, many restoration enthusiasts would like to purchase this original part. (Photo courtesy Mary Pozzi)

something non-absorbent to protect your paint. Use line wrenches to disconnect the brake lines from the master cylinder. Remove the nuts from the master cylinder and pull it off the booster.

Proportioning Valve

In order to balance the front brakes with the rear brakes, you can install an adjustable brake-pressure valve. It goes between the master cylinder and the rear brake system. The valve limits the amount of fluid going to the rear brakes in an attempt to keep the rear brakes from overpowering the front brakes. This could cause the rear brakes to lock up before the front brakes, causing loss of control of your car when slowing, especially during a panic stop.

Residual Pressure Valve

This valve is designed to work on rear-drum-brake vehicles. Rear-disc-

25 Baer suggests removing the pressure regulator valve from the side of the subframe in the rear brakeline and the barrel-shaped hold-off valve located below the master cylinder. If you're installing both the Baer front and rear disc brakes, these valves could adversely affect the disc-brake operation. If you want to leave these valves in place for aesthetics and so you don't need to change the main brakelines, simply disassemble the valves and remove the guts (piston, spring, moving parts, etc.). Don't remove or tamper with the rectangular distribution block and safety switch located under the master cylinder! (Photo courtesy Mary Pozzi)

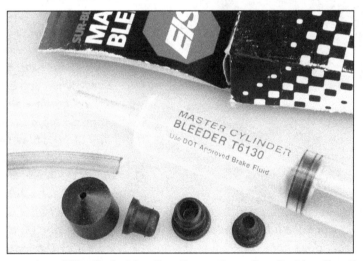

26 Start bleeding brakes at the wheel cylinder or caliper farthest from the master cylinder. Suction bleeders don't always work because they draw air bubbles through the threads in the bleeder screw, causing difficulty in bleeding the system. Gravity-bleeding the system by opening the bleeder screw and allowing gravity to push the bubbles out of each corner of the system works well if the system is primed; it's not a good idea to leave the system open for very long because it allows moisture to get in the system. Rather, you can have a buddy pump the pedal, open the bleeder, and then close the bleeder. Repeat until all bubbles have left the system. We found a much cleaner and easier way to bleed brakes. Danny Nix at Classic Performance Products (CPP) suggested we try using the syringe method. Before starting to bleed the system you need to double check the whole system and make sure that all the fittings and connections are tight. A bleeder kit like this is a cost-effective system. (Photo courtesy Mary Pozzi)

Disc Brake Installation CONTINUED

The forward master-cylinder reservoir on first-generation Camaros holds the fluid for the front brakes; the rear reservoir is for the rear brakes. The Baer kit has a new CPP master cylinder with the ports in the correct locations. This allowed us to leave the expansion loops in the brakelines and to leave the distribution block in the stock location. Baer includes adapter flare fittings to make it easier to adapt new lines due to different line sizes between manual and power brake systems. We were tempted to use our original master because it was a 1-inch bore, the same as the Baer master. The project car was a driver, not a concours restoration, so it made sense for us to use their supplied unit and not take any chances with the brakes. (Photo courtesy Mary Pozzi)

27 Make sure the farthest bleeder screw is open, a clear plastic/nylon hose with one end connects to the bleeder. The other end should be draped in a clear plastic or glass container with the end of the hose under about 1/2 inch of fluid (to keep air from possibly moving up the tube). Put the snubbed rubber tip on the end of the syringe in the port of the bottom of correct reservoir and push the plunger down to force the fluid into the port. Make sure you don't force air into the system, or else you'll have to start the process over again. Depress the plunger slowly when you start so you know how much fluid to add. Have a friend watch the hose to tell you when the bubbles stop coming out. Keep adding more fluid with the syringe until the bubbles stop. Don't allow the reservoir to empty or overfill during this process. Once the air stops coming out of the hose, close the bleeder screw and move to the next bleeder screw and continue until all four corners are free of air bubbles. Fill the master cylinder to about a 1/4 inch from the top and close the lid completely. Pump the brake pedal and hold pressure on it to confirm that it feels good and that the pedal does not slowly fall to the floor. It should stay firm and not move. If it moves, there's a loose connection somewhere in the system and it needs immediate attention. If everything checks out well, put all the wheels back on the car and torque them to the proper spec for your size wheel studs (stock: 7/16 inch, 75 ft-lbs; aftermarket: 1/2 inch, 80 ft-lbs.) (Photo courtesy Mary Pozzi)

brake-equipped cars do not typically use a residual pressure valve. The residual pressure valve holds a few pounds of brake fluid pressure in the rear drum brake system when the brake pedal is released, and the pressure in the rear system drops. The pressure held in the rear system keeps the rear brake shoes partially actuated. The shoes do not stay engaged against the inside of the brake drum; rather, they drop slightly away from the drum. Without the residual pressure kept in the system, the shoes would retract farther away from the drum and greatly increase rear brake reaction time when the brake pedal is pressed. The residual pressure also helps the rear drum self-adjustment system operate correctly.

CHAPTER 9

FRONT SUSPENSION

You can tell a lot about the condition of your Camaro by inspecting certain parts. Many of the suspension pieces and the alignment of the components tell a story of the life your car has led up to its current state. It's important to perform a detailed inspection before removing parts. You will be flying blind and be surprised with problems after assembly if you don't perform these checks ahead of time. If you follow the steps in this chapter, you'll have the easiest sequence of disassembly and you won't miss any important steps to preserve key data.

Take your car to a shop with an alignment rack and the proper frame equipment to quickly and accurately confirm that your car is straight. If you're aware of a previous accident or suspect your chassis is bent, a professional check is well worth the time and money. (Photo courtesy Steven Rupp)

Pre-Disassembly Inspection

Examining the alignment shims in the upper control arm mounts is the easiest way to determine the condition of your subframe, control arms, and body. If one side has a significant number of shims compared to the other, it's a good indicator of some sort of previous trauma. For instance, an alignment shop had to compensate for some sort of accident, which could have knocked the frame out of alignment, bent the frame, and/or bent the control arms.

The least expensive problem is severely deteriorated body bushings and body-to-frame alignment, which can be fixed by replacing the bushings and re-aligning the frame to the body.

A bent control arm can be difficult to recognize without the arm off the car and sitting right next to another good arm on your workbench. If your frame or body is bent enough that an alignment shop had to compensate, you've got some serious work ahead of you that should be done by a frame alignment shop. If it's only the frame, you can always purchase a replacement frame from a wrecking yard or an aftermarket frame from sources, such as Detroit Speed and many others.

To check the alignment of your frame to the body, simply take a peek under the car to confirm the alignment holes in the center frame mounts line up with the corresponding holes in the body right above them, next to the body bushings. Even if these holes line up, that doesn't rule out a bent frame, but enough force from an accident typically shifts the frame a little bit.

CHAPTER 9

This original rubber body bushing has seen better days. The one between the frame and the floorpan appears to be in decent shape but the lower half is in bad shape. The top of the frame is also slightly bowed from years of abuse.

Wrap your hand around the joint and linkage while a buddy rotates the steering wheel back and forth about 1 inch. If you feel clunking or discernable free-play it's time to replace the component.

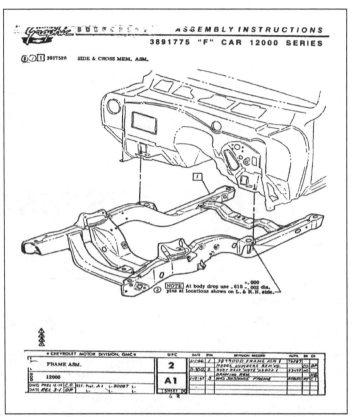

This diagram shows the alignment holes in the frame that match up with the holes in the body at the base of the firewall. If your car has been in an accident, these holes may not be the ideal alignment, which is covered later in this chapter. (This image is reprinted with the express consent of Year One, Inc, a licensee of General Motors Service Operations)

To further check the frame for straightness, you can take your car to a frame alignment shop and have it perform all the checks with its high-dollar equipment, or you can spend time with plumb-bobs, string, a completely flat and level surface such as a garage or shop floor, and the alignment specs listed in the Fisher service manuals, which are available at Year One and other parts suppliers. Always take time and common sense into account before spending many hours performing all the checks yourself—if you find a bent frame, how are you going to straighten it at home?

Your time and money may be better spent on having a shop equipped with the tools to fix your frame if it finds a problem; you can spend your time working on another part of your project. Before a frame alignment shop is able to determine and/or fix a bent chassis, you're going to need good body bushings. Aligning a frame with bad worn-out body bushings is like throwing your money in the trash.

Check the condition of your suspension components before you start removing any parts. There's no reason to replace a perfectly good part. For about 98 percent of the cars being considered for a restoration, you'll probably end up replacing most of the wearing parts, but there are exceptions. At least being armed with the following information, you can make your own judgment.

Visually inspect the six rubber bushings holding the frame to the floorpan and radiator support for cracking, deterioration, and abnormal bulging. Check the upper control arm bushings for the same signs of problems that would require replacement. Inspect the rag joint between the steering column and the steering box for cracking and deterioration.

Separation Anxiety

In order to perform a true "frame-off restoration," you must complete the big task of separating the subframe from the body. A common misconception is that you need to have a lift in your garage to separate the two. Using a 4-post lift does simplify the process, but it's not a requirement. The following is a way to perform this step with some common materials you can purchase or may have sitting around your garage.

FRONT SUSPENSION

You don't need a lift to separate the body and frame. You simply need jack stands, some other sturdy supports, and a couple of floor jacks for lowering the frame. Take the engine and transmission out, disconnect everything, and drop the frame. Be safe and don't take chances. For safety, use additional supports before climbing under the car.

Some body bolts and cage nuts can be very problematic to remove because years of rust have fused the nuts and bolts together. In particular, the body bolts that thread into the cage nuts at the base of the firewall and under the front seats tend to be difficult to remove. If you don't add penetrating oil before loosening the bolts, the nuts can break the cage and create a lot of extra work. Before lifting the car on the blocks, loosen all the body bolts a little, so you're not torquing the body and frame a lot while it's supported in the air.

This step can be performed with a complete Camaro. You simply drop the subframe out the bottom of the front clip, so you can rebuild the frame, or you strip the car down to just a frame and body shell. It's much safer to do this process without the engine and transmission because you can handle the subframe much easier. I've seen the following process done with a complete car by lifting the engine and transmission up in the engine compartment before dropping the frame out the bottom. If the front sheetmetal is still attached to the body shell, you must remove the wheels, tires, and all the front suspension before the frame will slide out from under the car.

Make sure all the parts are bolted to the body and frame. For instance, the fuel lines and brake lines are bolted to the body and the side of the subframe. Remove or unbolt all the shared components that link the two together. The brake lines under the master cylinder and combination valve are attached to the subframe and the front bumper and its brackets. These brackets are attached to the front of the frame. The emergency-brake cables are attached to the subframe hooks and the intermediate cables on the body, and they run through the driver's-side frame rail. The steering box is attached to the steering column with the "rag joint." The engine connects to the body through the exhaust, heater hoses, electrical wiring, ground straps, and transmission and engine mounts, which should be disconnected.

The subframe is bolted to the body in six places. Two bolts mount to the radiator support with nuts; two bolts thread into two "cage nuts" captured in the mounting pad in the lower part of the firewall; and the last two bolt the subframe to two cage nuts which are located directly under the front seats.

In cases where the car has a substantial amount of rust, the body bolts seize to the cage nuts and no amount of WD-40 or penetrating oil will loosen the bolts from these cage nuts. In fact, if you strip the cage nuts from the cage that keeps them in place, you have to cut out the cage and cut the nut loose from the body bolt. It's major surgery and quite a mess, so this is something you want to avoid.

Going through the access holes in the front side of the firewall is the easiest way to get to the cage nuts in the lower firewall, but you may have to remove the inner fenderwells so you can easily work on the cage nuts. The rear-most subframe bolts that are threaded into the cage nuts under the front seats can also break loose and cause problems; these cage nuts are much harder to work on because they are located under the seat frame-support panels. If a rear cage nut breaks loose, you have to perform some major surgery to repair it. Therefore, before attempting to remove these two bolts, it's a wise precaution to

HOW TO RESTORE YOUR CAMARO 1967–1969

remove the front seats, pull the carpet back, and squirt some penetrating oil on the tops of the body bolt threads and the cage nut.

Jack up the rear to allow enough space to work, clean, or paint the floorpan. Even if a floorpan is in good shape, it needs a thorough cleaning and some new rust-preventive paint. (This treatment provides many more years of protection against moisture and other deteriorating elements, and it doesn't require more than 20 inches of clearance to do the job.)

Support the rear frame with sturdy blocks of wood or jack stands. The car shown (on page 113) still has its rear axle, so we lowered the rear axle onto the jack stands. Then we moved the floor jack to the front of the subframe and jacked up the frame, which raised the body high enough to slide large sturdy blocks of wood with a couple of 2x6s (about the length of the rocker panel) to distribute the body weight across most of the length of the rocker panels. We positioned the 2x6s against the pinch weld on the bottom of the rockers. The front of the boards are about 1 inch forward of the front of the rocker, to distribute the weight on the strongest and flattest parts of the rockers.

Then, we lowered the car onto the wood and confirmed that nothing shifted. Once the body is supported on the wood, stop lowering the jack. You don't want to let the jack all the way down because the extra weight of the tires, wheels, and suspension mounted on the subframe is enough to cause the body to tilt forward and slide off the wood, which you want to avoid. We're trying to keep the body level on the wood and drop the frame out from under it.

You need a second floor jack placed under the transmission crossmember to hold up the rear of the frame while you remove the bolts. Remove the center body bolts located at the base of the firewall and then remove to the rearmost body bolts. Carefully lower both jacks at the same time to lower the subframe evenly—you've accomplished the "frame-off" portion of the job.

If your car is in the same configuration as the Camaro shown, the frame can be rolled out very easily because applying a little downward force on the front of the frame raises the rear of the frame.

Front Suspension Tools

Whether you're removing the front suspension from a complete car or a bare frame, you're going to need some specialty tools, such as brake spring pliers, a brake spring washer tool, an internal coil spring compressor, and ball joint removal tools. Pickle forks/ball joint separators are decent tools if you are going to replace the ball joint. The most elegant tool you can use to break ball joints loose is a pitman arm puller. They come in a couple different sizes, so they can be used for the ball joints and the tie rods as well as the pitman arm. Used properly, they don't damage the components or rubber dust boots.

You also need an internal coil spring compressor to safely remove and install the front coil springs. (Spring compressors with external fingers will not fit because of the limited space). You can rent these spring compressors from some auto parts chain stores, or you can purchase one to have around for future projects.

Front Suspension Removal

To completely remove the front suspension, you need to follow these procedures:

First, disconnect the brake line, being sure not to get brake fluid on painted surfaces. It's much easier to disassemble all the brake parts from the spindles before removing the spindle from the control arms (if you're going to replace the spindle and drum brakes for a new spindle and disc brakes, you can leave all the brake parts together).

If you have drum brakes slide the brake drum off the hub and get out your camera or a pad of paper and a pencil to take notes of where the springs and the brake hardware are attached. You can remove all the springs and parts without having the right tools, but you are much better off spending a few bucks now because you need brake spring pliers and a brake spring washer tool if you are going to re-install your drum brakes.

Remove the dust cap, cotter key, and large nut. Then pull the hub off the spindle, but be careful or the outer wheel bearing will fall on the ground and get dirty. Expect to see a lot of grease. Pop out the inner wheel bearing and stick both bearings, large nut, washer, and cotter pin in a plastic bag for later inspection. Put a plastic bag over the spindle pin or clean it off, or the grease will get all over the place during the rest of the process. Remove the backing plate from the spindle.

If you have disc brakes, remove the brake caliper first, then remove the dust cap, cotter pin, and large nut. Slide the hub off the spindle, but don't allow the outer wheel bearing to fall off because it makes a mess. Pull the inner bearing off the spindle and place both bearings in a plastic bag with the nut, washer, and cotter pin for later inspection. Place a plastic bag over the spindle pin or clean off the grease or it'll get all over the place during the rest of the process.

Loosen the large nut that holds the pitman arm on the steering box, but don't remove the pitman arm yet. Use your pitman arm tool to remove the outer tie-rod ends and then remove the outer tie-rod end from the steering arms on the spindles.

Now use the same process to remove the inner tie rod ends from the center link/drag link, then separate the drag link from the pitman arm and idler arm. Remove the last bolt holding the steering arms to the spindles and remove the arms. Remove the caliper bracket and backing plate from the spindle. Remove the idler arm from the frame and the pitman arm from the steering box.

If you haven't already done so and have power steering, remove the power steering lines and drain the power steering fluid into a container for recycling. Unbolt the steering box from the frame. Be careful during this process because it weighs about 50 pounds (a little lighter for manual steering boxes), so you need to support it and be careful not to let it fall on you because it can maim you, or worse, if you're under it.

Remove the brake line tabs from the side of the frame and put them in a bag marked "right" and "left." Now remove the sway bar end links, then the sway bar. Unbolt the shocks and slide them out the bottom of the control arm. Slide the internal spring compressor through the lower control arm where you removed the shock from. Install it on the compressor and compress the spring enough to where the spring is a little loose.

Place your floor jack under the lower control arm and compress the arm a little bit. Remove the cotter pins from the upper and lower ball joint and loosen both nuts a few threads, but don't remove remove them yet. Use your larger pitman arm tool to break the lower ball joint loose from the spindles, and then use your tool to break the upper ball joint loose from the spindle. Since the jack is holding up the lower control arm, you can remove the lower ball joint nut.

Now you can let the jack down slowly, making sure the coil spring doesn't pop out, which it shouldn't do if you used the right tool. If you didn't compress the spring, it can fly out of the lower control arm when you let the jack down and possibly maim somebody. The compressed spring should fall out onto the ground as you lower the jack.

Perform the same process to the other control arm.

Remove the upper ball joint nuts and the spindles. Take a picture and note all the alignment shims between the subframe and upper control arm shaft. Loosen the nuts and pull the shims out. Wrap some tape around the front pack and then the rear pack to keep a record of what was in the car. Then place them in a plastic bag marked which side of the car they were removed from. These are good notes for later assembly and the alignment shims also tell a story about possible bent frame or control arms from a previous accident.

Remove the nuts holding the upper control arm shafts to the subframe. If the engine is still in the car, knock out the knurled bolts that are pressed into the frame tower in order to get the control arms off the subframe.

Remove the lower control arms. Get your deepwell socket, extension, and ratchet and stick it through the access hole in the front of the frame. Remove the nuts on the rear-most lower control arm bolts. Be careful not to drop the bolts or your socket and extension into the frame. Remove the nuts from the forward lower control arm bolts. Pull the rear-most bolts out of the frame. Use a hammer and a brass drift to tap the bolts out, if they won't easily come out of the frame.

Now remove the forward bolts and drop the lower control arms out of the frame.

If you're going to perform a full-restoration of the subframe and the engine is out, remove the engine "stands" from the frame. There are access holes in the bottom side of the frame to get to the back side of the bolts. Clips that hold the front brake line are bolted to the subframe in multiple locations, which are easily removed.

If your car was equipped with a manual transmission, you'll have to remove the clutch linkage bracket on the outside of the frame. Depending on the year of your frame, there may be some additional brackets attached to the frame that you need to remove.

Subframe Changes

Chevrolet changed the subframes a little bit each year from 1967 to 1969. In 1967, the lower control arm didn't have a rubber bump-stop to limit compression of the suspension.

CHAPTER 9

Instead, the bump stop was located on the subframe, and it was designed to contact the top of the rear side of the control arm. In 1968 the bumpstop was moved to the top of the rear side of the control arm and a contact pad was welded to the subframe. In 1969 the bump stop was moved to the front side of the control arm, and the contact pad was moved to the front side of the coil spring pocket.

The subframe had other design changes over the course of the three years of the first-generation F-body. One of those changes is to the center body-bolt towers that attach the frame to the bottom of the firewall. (I understand why Chevrolet changed the design after personally seeing the deficiencies of the structural strength when an early 1967 subframe was subjected to big-block power.)

The 1967 frame tower was welded to the subframe's outside face and was not as strong as the redesigned version of the tower. The revised tower featured a gusset welded to the top of the subframe and attached to the early 1967 tower. Some time in 1968, the mounting tower was changed again and the downward leg to the side of the frame was deleted, but the new tower was still welded to the top of the frame. My personal opinion is that the first revision was probably the strongest design due to triangulation.

Another noticeable subframe change is the number of mounting holes on the top of the front frame rails where the radiator support mounts meet the frame. The 1967 subframe had two holes on the top of each rail, and the 1968 and 1969 units had three holes on each side. In 1968, Chevrolet had completely redesigned the Nova and used the same subframe as the Camaro and Firebird. The new center hole was where the Nova radiator support mounted. Mid-year in 1967 the factory added a large, round access hole to the inside face of the front frame rails.

Subframe and Control Arm Problems

The following is a list of common problems with subframes and control arms that you should be aware of while performing a restoration or considering purchasing a car or used parts for your own Camaro project. Educating yourself with these issues helps you identify possible problems that may or may not be obvious.

Mount Holes

It's a good idea to separate the subframe from the body since it allows access to the top of the frame for inspection of the welds and condition of the metal without everything getting in the way. Subframes suffer from rust around the body-bushing mount locations because after years of movement the bushing removes the paint and leaves an exposed metal surface. Therefore, it's not uncommon to have serious rust around the six body-bolt mount locations. A few companies offer subframe repair plates, so you can cut out the damaged metal and weld in new sections.

If you're a good fabricator, you can make your own plates and weld them in. Keep in mind that you'll need to cut out the thin rusted metal before the new thicker metal can be welded to the frame.

Center Body Mounts

Of the three years, the 1967 unit was the most notorious for breaking the center body mounts. There was a change in the front frame horn design mid-way through 1967, but it's not known if the center tower was changed at the same time. If it was, then the problematic design was an early 1967 problem. Either way, the tower was only welded to the outside face of the frame and mixed with the poor factory welds made the design susceptible to cracking at the welds. If you are going for an all-original restoration you can weld up the problem welds, and if you don't mind improving the design you can add a gusset to replicate the later design for added strength. These problems can be repaired with some good fabrication skills.

Upper Control Arm Tower

The stock subframe was never designed to be put through 40-plus years of abuse and therefore often needs to be repaired and sometimes replaced. The area that gets the most abuse is the upper control arm mounting tower.

Flexing and jarring from years and thousands of hard-driving miles the upper control arm tower starts to show signs of fatigue and starts to crack in a few common areas. It doesn't help that the factory welds left a lot to be desired, and fractures have been found on the top of the frame around the perimeter of the welds of the control arm tower.

The most common cracks are found in the weld between the top of the frame and the base of the shock mount plate. Humans (not computer-controlled robots) welded the the factory subframe, so the welding is not precise or clean and differed from frame to frame. Some welds burned through the metal and left very little material to rely on for strength.

FRONT SUSPENSION

A good fabricator and welder can drill the ends of the fractures to keep them from spreading and weld the problem areas. With some skill, this can be done with very little evidence.

Sway Bar Mount

The sway bar frame brackets are fastened to the underside of the frame rails using threads tapped in the steel. In some cases, the threads simply strip out from rust corrosion, and in other cases, the threads pull out. If this happens, you can drill out the hole and weld a nut up inside the frame, or you can use a long wrench to reach inside the frame rail to hold a nut in place while tightening the sway bar mount bolts.

When using larger-than-stock sway bars and driving aggressively, the frame often becomes fatigued from constant flex, and as a result, the metal around the mount simply cracks and the frame pulls apart. You should weld-in gusset plates to reinforce this critical area.

Lower Control Arm Holes

Inspect the lower control arm bolt holes. They should be round, and the bolt should not have excessive play. Accidents, deep pot holes, and striking debris at high speeds can impact the lower control arms and cause the lower bolt holes to distort and become more of an oval than a round hole. If this has happened to your frame, a good fabricator can drill out the frame and essentially weld a hardened washer into the frame.

Lower Ball Joint Pocket

Although the control arms aren't really part of the subframe, they do have problems. If the ball joints have been replaced, the ball joint pocket in the lower control arm can be distorted/stretched. The ball joint needs to be tight on the control arm. We've seen people use poor judgment and weld the ball joint to the control arm because it was loose. The fix for this problem is to ensure you're using the correct ball joint, have a shop shrink or repair the ball joint hole, or replace the control arm.

Control Arm

Because the front leading edge of the lower control arm hangs down, it often comes in contact with road debris and cement parking blocks. As a result, the arm is bent from the contact. Just because the leading edge is bent, doesn't mean that the arm is bent beyond its specifications keeping it from doing its job. If the rest of the control arm doesn't look distorted, you may be able to compare it to your other control arm or another Camaro control arm. A competent mechanic can repair a bent leading edge with a vise and a good hammer.

Subframe Seams

In some cases, a couple of factory welds didn't get good penetration, which can be touched up by a good fabricator. If you're considering welding up every seam on the frame, be careful not to put much heat in the frame or you'll end up with a warped subframe.

Don't start on one end of the frame and weld the whole thing in one long bead. Take measurements from corner to corner in an X-pattern between the centerlines of the body bushing holes and lay the frame on four jack stands. This gives you an idea of how straight your subframe is before welding it. Stitch-weld the frame about 2 or 3 inches at a time and alternate from side to side, so you don't put a bunch of heat in one area at one time.

Also consider that welding all of the subframe seams detracts from the originality of the car and once you start, there's no turning back to stock.

Painting Suspension Parts

Please take into account that only a few parts on a stock Camaro are painted gloss black and none of those parts are the subframe or any part connected to the subframe. I think nothing looks more horrid than a Camaro pretending to look stock yet has a glossy black subframe or suspension parts.

Another common mistake made in Camaro restorations is the use of one color to paint all the parts; for instance, the tie-rod ends, adjustment sleeves, and all the other steering linkage were not the same color. Painting all the steering components the same gray cast-coat paint looks hokey. The suspension parts are all different colors because they all came from different parts suppliers. In fact, some of the parts weren't even painted from the factory, and they were made from different materials. This made them different colors by default, but if you left components unprotected without paint they would rust and look horrible.

For greater effect, take a few extra steps with different cast-coat colors and different shades of black paint to make your restoration look a little more correct. Also, consider that black isn't just black. Professional restorers recognize that there were different shades of blacks, including bluish-blacks, redish-blacks, brownish-blacks, etc. There are also different black finishes such as flat black, semi-gloss black, matte black, etc.

If you're going for a concours restoration to win shows, educate yourself on which colors and finishes came on the components corresponding to the year and model of your Camaro. It would also help to educate yourself on details that may have differed between the Norwood and Van Nuys Camaro plants. To start with, the subframe itself was semi-gloss black from both plants. Some rust-resistant paint companies (like KBS Coatings) make a durable semi-gloss black product that's not exactly like the stock paint but it's durable. The KBS topcoat is called Blacktop Chassiscoater, which goes over its rust preventative coating named Rust Seal. Eastwood and OER make multiple gray cast-coat colors that can be used in order to have different shades of gray to mimic factory restorations.

Getting just the right color and finish is not as easy as using a few spray cans. But even if you don't have years of knowledge of finishes and how to produce them, you can now produce a much nicer-appearing restoration compared to the typical shop or hobbyist that sprays everything one color.

Rebuilding the Subframe

This can be done before or after bolting the subframe to the body. We decided to rebuild the subframe prior to mating it to the body. When you have all your parts painted, plated, and coated you'll have to reassemble it.

Here is a brief overview of the procedures for reassembling the subframe:

Set the subframe on four sturdy jack stands. You can use rags between the stands and the frame if you want to protect the finish.

Bolt the control arms on the frame without installing any alignment shims in the upper control arms.

Use your spring compressor to install the front coil springs and

This 1967 received a concours restoration from Super Car Workshop using known factory finishes along with original inspection marks found under layers of grease—now restored back to their original luster. The factory labels have also been reproduced by Super Car Workshop with the correct font, which isn't always the case with labels available to the general public. Instead of oiling bare metal to keep it from rusting, many paints can resemble the correct metal finish. (Photo courtesy David Boland)

This 1969 ZL1 was restored by Mark Schwartz back to original condition using knowledge of factory finishes and inspection marks. From this angle you can tell that the steering linkage parts aren't all the same color. Some restorers paint everything the same color of cast-coat gray not knowing or caring that it's not correct. These parts came from different sources and were different bare materials, so they would not be the same colors.

assemble the spindles on the upper and lower ball joints.

Bolt the steering box to the frame. Install the idler arm, pitman arm, drag link, and all the tie rods ends.

Install the front sway bar now, or you could have problems installing it after you bolt the brake rotors in place, or you'll scratch all of your good work.

Assemble the front brakes and the front brake lines.

Now you're pretty much ready to bolt the frame to the car. The rest of the parts can be assembled after the frame is bolted to the body.

Replacing Body Bushings

If you're performing a correct restoration, get some new rubber body-isolator bushings. If you're going to add polyurethane body bushings you may want to get a new bushing kit like the one from Prothane.

If you plan on installing subframe connectors now or in the future, install solid body bushings from Detroit Speed. The rubber and polyurethane bushings allow too much movement of the frame and defeat the purpose of installing subframe connectors and can cause serious frame fatigue. Detroit Speed offers standard (same height as stock bushings) and half-height bushings, which are shorter than the stock bushings and lower the body on the frame and the ride height. Drawbacks to the shorter bushings are the reduced clearance between the body and the transmission and the increased difficulty of installing the transmission crossmember. It also changes the steering column angle and moves the engine a little higher, reducing hood clearance by about 1/2 inch.

Installing the Subframe

Hopefully you've already had the frame removed from the body to fully clean and paint the subframe and the underside of the body. Use extreme caution to not have the frame or body fall on you during this process. The body should be propped up with safe stands so it won't tip, slide, or fall while installing the frame.

Before installing the subframe, install the full-length brake line and fuel line up against the body because it's much easier to get it in now instead of trying to fish it through the gap between the floorpan and the frame.

In order to install the subframe you do the reverse of taking it out; just be careful with all your new paint. Be extremely cautious not to smash the brake and fuel lines. Make sure you drop the four floorpan body bushings into the proper holes in the subframe. If you have rubber or polyurethane body bushings, the largest part of the bushings with the metal faces upward and rides against the floorpan. Install as the accompanying kit instructions state.

Have the bottom half of your bushings ready to install (if they are a tight fit to the upper bushings, go ahead and assemble them). All types of body bushings install dry, unless otherwise instructed by kit directions.

Once the frame is in position and all the parts and hardware of the body bushings are in the proper sequence, you can temporarily align the frame to the body. There are two rear body bushings and two center body bushings that mount to the body at the base of the firewall. The frame pads for those center body bushings have corresponding alignment holes in the body at the base of the firewall. As long as your frame is straight and the body shell has not been wrecked, the holes should be in good alignment. If there's any hint of possible trauma to the firewall area or frame these holes may not line up correctly. Temporarily align the frame to these holes as long as you confirm it later after the car is completely assembled and you're ready to set up the wheel alignment.

If there's a question of possible frame alignment issues, set the toe to zero by eyeing it before adding a single alignment shim. Use a plumb-bob, center the tires in the fenderwells, measure from the center of the rear wheels to the center of the front wheels, check side-to-side positioning, and adjust the frame to the correct location. To align the frame holes you can stick a long ratchet extension or pry bar

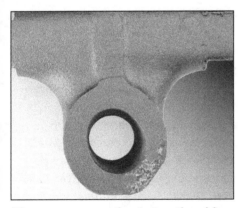

It's common practice to tap the sides of the spindle knuckles with a hammer to break the ball joints loose. This action is bad for the spindle because it can distort the cast iron. The distortion shown here is from driving with a loose lower ball joint.

through the holes and move the frame into position.

Once the car is completely assembled, there will be a lot of tension on the body bolts, which makes them difficult to adjust. Don't forget that if you're doing this after the car is assembled, the two front body bushings at the radiator support must also be loose. The four main body bushing bolts' final torque is 85 ft-lbs and the radiator support bushing nuts are to be torqued to 40 ft-lbs.

If you're really creative, you can install the engine and transmission onto the subframe, but the additional weight adds a whole new level of difficulty. This isn't suggested unless you're working with an actual hydraulic lift to raise and lower the body onto the frame. It's easy to accidentally smash the engine into the firewall and damage both in the process. Don't take chances and be very careful.

Subframe Installation

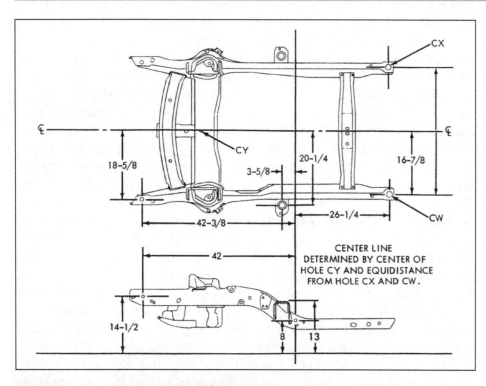

1 *For the dimensions of your front subframe, you'll find helpful photos diagrams like this in Chevrolet Chassis Service Manual and for more detailed specs of frame-to-body alignment pick up a Body by Fisher Manuals for the specific year of your Camaro. (Image reprinted with the express consent of Year One Inc, a licensee of General Motors Service Operations)*

2 *When restoring the suspension back to original parts you can search swap meets and online auctions for NOS parts. For our projects we aren't keeping originals we choose MOOG parts and so do suspension engineers like Kyle Tucker from Detroit Speed. Refinish the control arm before installing the bushings. This arm is shown before it was painted the correct semi-gloss black.*

Subframe Installation CONTINUED

 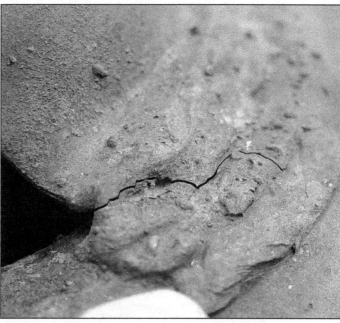

3 *The upper control arm tower and shock mount get a lot of abuse and they don't get a lot of attention. The shock plate can be cracked where it meets the frame rail and can go undetected because it's covered by the control arm. Hairline fractures commonly start around factory welds at the base of the control arm tower.*

4 *The center mount pedestals on the subframe changed a little over the three years to increase strength. The 1967 mount was weak so it was redesigned to mount to the top of the frame, as shown in this design.*

5 *The factory lower control arms changed each year to match the change in the subframe bump-stop locations. Starting in the 11 o'clock position and proceeding clockwise are the 1967 arm, the 1968 arm, and the 1969 arm. In 1967, the bump stop was located on the frame, and there was no hole in the control arm. The 1968 arm had the bump-stop on the rear of the arm, and in 1969 it moved to the front of the arm. Chevrolet sold a service replacement arm (shown at bottom) with holes for the 1969 and 1968 (arrow).*

CHAPTER 10

REAR SUSPENSION

The Camaro rear suspension is a weak point of the 1967 and 1968, as we'll explain little later on, but suffice it to say that the original mono-leaf rear springs and rear suspension setup does not provide adequate control and performance, particularly for a V-8-powered Camaro. Rather, the springs coil up and unload, causing severe wheel-hop and unsafe handling characteristics. Safety is paramount when it comes to driving any car, and therefore the mono-leaf springs should be replaced unless you're restoring the Camaro back to factory original equipment. There are a plethora of aftermarket as well as OEM multi-leaf spring upgrades that make performance far better than the original setup and do not adversely affect the value of a street-driven car.

1967

By 1967, Chevrolet had introduced the Camaro, and the Bow Tie group finally had a vehicle to compete for a share of the "pony" car market. As with many first-year vehicles, not all the bugs had been worked out of the platform, and the Camaro had

The original shocks are referred to as spriral shocks because of their distinctive spiral bead stamped in the body. These shocks are highly sought after by restorers. The original finish was gray semi-gloss.

rear suspension problems. The rear suspension was comprised of two single-leaf springs and two shock absorbers, which Chevrolet called the "mono-plates." These springs have been commonly referred to as "mono-leaf" rear springs.

While these springs were ideal for 140- or 155-hp 6-cylinder-powered Camaros, they lacked the spring rate to cope with aggressively driven V-8-equipped Camaros. Under power, the springs wrapped-up, which meant they twisted into an "S" shape instead of staying in their arched shape. To make matters worse, both rear shock absorbers were located in front of the rear axle, which was a poor location. Consequently, under hard acceleration, V-8 Camaros experienced severe wheel hop because both rear shocks rebounded and compressed on the same arc while the rear mono-leaf wrapped-up. As the spring twisted into an "S" shape, they would eventually stop resisting twist causing loss of tire traction for a split second until the springs could unwind. The tires would get traction then the wrap-up would start again.

The shock location design allowed the rear axle to easily hop straight up and down as the tire was gaining and losing traction. The wheel hopping would get violent and unsafe, causing the car to be uncontrollable and causing many accidents when unsuspecting drivers lost control of the vehicle.

About halfway through the 1967 Camaro production run (sometime in December), Chevrolet added a link/bar that connected the rear axle housing to the side of the frame and floorboard under the passenger-side rear seat. This bar was called a "radius rod" and was designed to help inhibit wheel hop. The radius rod was not installed on all V-8 Camaros and had three designs—the first design of the bar was round, the second design was square with an additional travel-limiting bracket that was welded to the rear axle housing, and the third design was a service upgrade radius-rod assembly to upgrade Camaros not originally equipped with the radius rod or still equipped with the older round radius rod. The service upgrade radius rod is easier to spot since there were no factory bolt locations on the side of the frame or under the rear seat, which required the bracket to be welded directly to the frame and floor instead of being bolted into place.

Perform a thorough inspection to determine whether the shocks, sway-bar bushings, ball joints, control-arm pivot bushings are in good working order. If they are not, you need to refer to a service manual for replacement because the suspension system is a safety system and it needs to be in good working order.

1968

Starting with the 1968 model year, the Camaro's rear shocks were relocated into a staggered design position in which one shock was mounted in front of the axle and one was mounted behind it. This was a huge improvement from the previous year's non-staggered design in which both rear shocks attached to the rear axle in front of axle.

The change in shock location greatly reduced wheel-hop due to the compression and rebound of the suspension working on opposing planes. Chevrolet also started offering multi-leaf rear leaf springs on higher-powered V-8 models (SS and Z28). The additional leaves in the springs helped reduce wind-up. The mono-leaf rear springs were still used in L6 and lower-output V-8 Camaros.

1969

The 1969 rear suspension was essentially the same as the 1968, so the mono-leaf rear springs were still standard equipment on L6 and the lower-output 307- and 327-equipped Camaros.

Mono-Leaf vs. Multi-Leaf

If you're going to drive your restored or rebuilt Camaro on a regular basis and plan on driving it aggressively, it should not be equipped with mono springs. I need to reiterate my concerns about mono-leaf springs, especially on 1967 Camaros. We've already mentioned in detail about wheel-hop associated with mono-leafs and why the (1967 especially) suspension configuration is the culprit, but I didn't mention safety. Chevy designed the radius rod for the 1967 because aggressive driving of a powerful 1967 Camaro and Firebird with non-staggered shocks mixed with mono-leaf springs often is very unsafe.

If you're not restoring a Yenko, or another rare Camaro that must be kept original, please consider upgrading to multi-leaf rear springs. I've seen a few 1967 Camaros get wrecked due to suspension deficiencies. I am completely convinced that the wrecks would have never happened if the Camaro had been equipped with a multi-leaf spring. Take your invested time, money, and safety into consideration when making choices about the parts you put on your car.

Spiral Shocks

Original shock absorbers have a large spiral indentation that runs around the outside of the shock. If you're lucky enough to still have a set on your car, you should hang on to them, no matter what condition they are in. These old spiral shocks are very rare, and there aren't any companies currently rebuilding the internal components or making good reproduction units. Typically, the first time the shocks were replaced (perhaps some 30 years ago) the mechanic simply threw them in the garbage. These shocks are usually not in great shape now that they're 40 years old, but worth restoring if you're performing a correct restoration. There are companies restoring these shocks, but they only repair the finish on the shock, not the shaft seals. You can purchase reproduction spiral shocks without the date codes for your car and send them to a shop that can correctly stamp them for you if you're concerned about the details.

Aftermarket Shocks

Inside every modern shock there are shims and a piston that control the compression and rebound of the shock absorber, which affects the comfort level in the car while driving over bumpy surfaces and also assists the sway bar and springs to control suspension when cornering. Stock shocks are not valved the same as performance shocks, especially on these old muscle cars. If you are building a street car to be driven on a regular basis and aren't worried about performance driving, you can usually get decent results out of KYB shocks. If you need your car to handle well while doing some spirited driving on the street or on the race track, you may want to consider installing performance shock absorbers, such as those from Koni or Bilstein.

Spring Perches

The leaf-spring perches on the rear axle housing are different for the mono and multi-leaf rear springs. The spring perch for a mono-leaf spring has a distinctive shallow pocket for the leaf spring and isolators to sit in compared to the deeper multi-leaf perch

The mono-leaf spring has an alignment pin that fits in the hole of the lower shock plate to keep the rear axle centered in the rear wheel well. Because there's not a pin or nut on the top, the spring perch pocket sits directly against the housing tube.

The multi-leaf spring has an alignment pin on the bottom, but unlike the mono-leaf there is a nut and stud on the top as well. This design required a hole in the top of the spring perch and clearance between the spring perch and the housing tube.

As with all aspects of your restoration, you need to determine your level of authenticity. Your leaf-spring options are: Have original leaf springs rebuilt, buy new reproduction leaf springs, or purchase aftermarket springs.

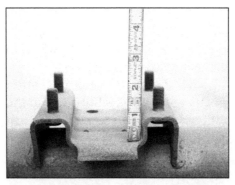

The mono-leaf spring perch only had to be deep enough to hold a single leaf and the rubber isolator cushions. Note the lack of the alignment hole in the center of the perch.

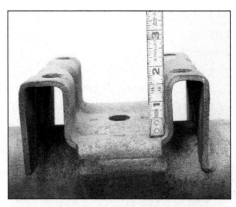

The multi-leaf spring perch is visibly much deeper for the larger spring pack. It has an alignment hole in the spring perch.

This 05D 1967 Yenko Camaro has the signature Yenko traction bars and the factory radius rod that Chevrolet installed post-midyear 1967 higher-performance Camaros. These were installed to reduce wheel hop. This square radius rod is a later version. (Photo courtesy of David Boland)

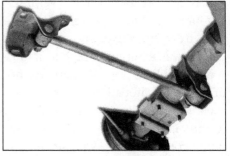

The early radius rod was round, and didn't have the additional travel-limiting bracket on the rear differential housing. Staggered-shock design and the introduction of multi-leaf springs on the 1968 model greatly reduced wheel-hop and the radius rod was eliminated.

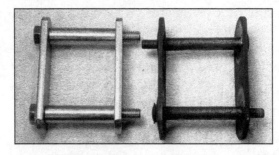

Before turning a wrench on the job of replacing rear leaf springs, you must purchase six new cage nuts and a new fuel-tank-to-filler-neck hose coupler (1967 and 1968). The reason

The stock shackle (right) is fairly strong. The factory finish on the shackle is semi-gloss black and some were gray phosphate. Prothane also offers a complete heavy-duty shackle and bushing kit. (Photo courtesy Mary Pozzi)

for the fuel tank coupler is that it's easier to replace the rear springs with the fuel tank out of the car and you can't reuse the old coupler because it will be too brittle.

REAR SUSPENSION

The original J-nuts or cage nuts are square and pressed into a spring steel clip. Six of these nuts hold the front leaf-spring bracket of the front eye of the rear leaf spring to the body. These cage nuts are usually rusted and the nut breaks loose from the clip, causing them to spin for eternity and cause hours of grief. If you add WD-40 after they are broken loose, the nut simply spins faster, so it's a futile battle.

Original J-nuts are not your friends. Even if you're working on a clean 40-plus-year-old Camaro that spent its whole life in the garage and never saw a drop of moisture or was well taken care of in one of the dry climates (Southern California, Nevada desert, most of Arizona, New Mexico, and most of Texas), you're still probably going to have trouble with your J-nuts. We strongly urge you to remove old leaf springs and to have six on hand before starting the job, especially if you're intending to make this a one-day job.

Leaf Spring Removal

By performing the following procedures, you will be able to remove the leaf springs.

If you're leaving the rear axle housing in place while replacing leaf springs, support it with jack stands in order to drop the springs out from the bottom. Don't forget to support the body separately from the differential housing and leaf springs.

Support the leaf springs on both ends. Unbolt the shocks from their upper mounts in the trunk. If your car still has the plastic factory caps over the shock stud sticking up in the trunk, you might have yourself nice original pieces. Pop them off and save them for later.

Unbolt the shocks from the lower shock plates. Detach the rear axle housing from the leaf springs by unbolting and removing the lower shock plates from it. If you're removing the rear axle housing, don't forget to pull the driveshaft out of the rearend yoke. Then, disconnect the brake line attaching the rear brake tee to the body.

Removing the front spring eye housings is usually the hardest part of disconnecting the rear leaf springs. There are three bolts that screw into J-nuts in the rear unibody on the front mount of each spring. Sometimes the J-nuts aren't a problem, but it's a higher probability they'll give you grief. Try to get some penetrating oil or WD-40 on the J-nuts up in the body before attempting to remove the bolts.

Rust causes the hardware to degrade over the years. It's not uncommon for one or all of the J-nuts to break in half, causing the bolt to spin but not come out of the J-nut. If this happens, you will need extra force. Remove as many of the three bolts as possible, leaving the bad one or ones until last.

Use a pry bar to pry the leaf spring away from the body to apply resistance and pressure to the hardware. Use an air-driven impact wrench to zap the bolts out of the nuts while prying. Some have said you can sometimes stick a screwdriver into the hole to pop out the nut that has come loose from the J-nut clip. If that doesn't work, you'll need to resort to cutting off the bolt heads with a die grinder.

Be sure you've supported the spring, so it doesn't fall on you when you get the bolts out. Mono-leaf springs can weigh about 45 pounds each and a multi-leaf spring can weigh as much as 75 pounds. Don't let the weight catch you by surprise and hurt you.

Remove the nuts from the through-bolts and the shackles and then pull the spring out.

Clean and paint all your parts.

Inspect the floorpan and the area around the front of the leaf spring. Camaros from wet and snowy climates have been known to have a lot of rust in this area. We've seen this area so deteriorated that the cage nuts were not held to the sheetmetal anymore. The previous owner had drilled through the bulkhead above this area and installed longer bolts with nuts located inside the car touching the rear seat base. It's a very bad idea because this area above the bracket is a hollow bulkhead that easily collapses and tears open.

These are reproduction J-nuts. They clip into the floorpan and hold the front bracket for the rear leaf spring in place. These nuts degrade over years of service. (photo courtesy Mary Pozzi)

Don't tighten the rear leaf-spring shackle bolts until the full weight of the car is on the suspension. This sounds strange, but if you tighten the shackle bolts with the weight off the rear suspension, the springs will be at full droop (full arc of the spring). In turn, the bushings will experience a certain amount of "stiction" that basically locks the bushings in that position and keeps the rear end lifted up with the springs still in the full-droop arc. As a result, the rear suspension sits at an exaggerated height. Of course, over time the bushings twist and distort causing the suspension to drop a little (this condition along with weakening of the spring is referred to as "settling"), but sometimes it can take years to fully settle down.

If you do tighten the springs with the springs loaded and the car still sits high, you may simply have improperly arched springs. Keep in mind that these cars came from the factory with a pretty high stock ride height.

You must swap more than just the springs when converting from a mono-leaf to a multi-leaf rear spring setup. Swap out the rear axle housing with one designed for multi-leaf springs, and purchase new lower shock plates.

The leaf-spring perches on mono- and multi-leaf-equipped rear-axle housings have a different depth for containing the different heights of the springs. The mono-leaf spring perch is only 7/8 inch deep and doesn't have a hole in the perch for aligning the leaf spring, but the lower shock plate does have a hole to keep the leaf spring in place. The multi-leaf spring perch has a deeper pocket (1¾ inches) to hold the multiple-spring pack. The multi-leaf perch also has the alignment pin hole in the center of the pocket. In addition, a hole in the correct multi-leaf lower shock plate allows the nut and alignment pin stud to fit through it.

What do you do if you're upgrading a mono-leaf car to a multi-leaf setup and still have the mono-leaf rear-axle housing? If you attempted to drill the necessary alignment hole in the housing spring perch you would drill right into the axle tube. You can't remove the bolt/alignment pin from the multi-leaf spring pack. Global West sells a kit (P/N 1035) that has spacers, shims, and hardware to allow installing the taller multi-leaf spring packs on mono-leaf spring perches. But it's best to get a rear axle housing with the correct perches.

Isolators In or Out

If you're installing original or correct restoration leaf springs in a car for the right application (i.e., multi-leaf springs with multi-leaf rear axle housing and lower shock plates), you should be able to use the leaf spring isolator pads on the top and the bottom of the leaf springs. If you're installing correct leaf springs for a factory-correct restoration, there should be enough space to install an isolator pad on the top and the bottom of each leaf spring and the lower shock plate. There should be no gap between the lower shock plate and the leaf spring perch on the axle housing with the leaf spring and isolator pads in place.

Some aftermarket leaf springs are thicker (height-wise) than stock springs and leave a gap between the perch and shock plate. This leaves you with the choice of removing the lower isolator pad to close the gap between the perch and shock plate or adding shims between them. With either solution, you should firmly lock the leaf spring in place between the perch and shock plate. If you have a gap between the shock plate and perch, the plate will start to bend and the hardware will start to bend and eventually fail.

The isolator pads are designed to stop road and driveline shock from being transferred to the suspension, chassis, and body, so the ride is more comfortable. Non-factory performance applications commonly run without the pads by either using different spring perches or swapping aluminum plates for the spring isolators. High-performance applications want to reduce as much cushion and flex as possible in the suspension so the car reacts more predictably, which is a much higher concern than soft-ride quality.

Leaf-Spring Bushing Types

When considering non-factory applications, you have the choice of leaf-spring eye bushings in rubber or polyurethane, depending on the level of ride quality you're after. Front bushings on the rear leaf springs transfer most of the road shock to the body, and that's why Kyle Tucker (ex-GM suspension engineer with a mechanical engineering degree from the University of Missouri-Rolla) at Detroit Speed prefers to use rubber bushings in the front spring eye. He states that it helps improve ride quality in daily-driver Camaros, but he prefers to use polyurethane in the rear spring eyes because the stiffer bushings and compression shackles help control suspension stability for performance driving. Mixing rubber in front for comfort and polyurethane in the rear of the leaf springs give you the

REAR SUSPENSION

If you decide to upgrade from rubber to polyurethane bushings, Prothane sells parts or complete kits to replace every suspension and body bushing. It offers many colors, but we chose black for a stock-style look. (Photo courtesy Randy Oldham)

best of both worlds for real-world streetable applications.

The 1967 lower shock plates were bolted to the leaf spring perch on the rear axle housing with T-bolts. To further strengthen the axle tube and keep it from separating from the spring perches in 1968 Camaros, Chevrolet started using U-bolts on higher-powered Camaros. The U-bolt went over the top of the axle housing tube to keep it in place on the outer axle-mounting points (T-bolts were still used on the inner attaching points). It's common practice for drag racers to weld steel plate gussets from the spring perch to the rear housing axle tube and use two U-bolts over the top of the housing tube on each side—they ditch the T-bolts altogether.

Performance Leaf Springs

A few companies offer performance leaf springs designed to give better performance than the original springs. The way they improve upon the original design is to give the leaf spring more front bias (increase spring thickness and strength as well as design secondary leaves to protrude farther toward the front spring eye). Good multi-leaves for street driving also have anti-friction pads under the ends of each secondary leaf to keep the leaves from binding against each other. Aftermarket companies, such as Hotchkis, Detroit Speed, and Global West, design their springs to lower the vehicle from 1¼ inches to 3 inches depending on the company. Eaton Detroit Spring makes stock springs to original specs and it also makes custom springs to the customers' specifications of design and ride height.

Installing Leaf Springs

Our project Camaro for leaf-spring replacement is a 1968 daily-driven street Camaro that is not a concours restoration. It will be receiving a Hotchkis leaf-spring kit, which comes with leaf springs, heavy-duty shackles, bolts, cage nuts, polyurethane bushings, spring isolator pads, and U-bolts. It basically includes everything you need for this job, except for the tools and the vehicle. These springs lower the ride height of the vehicle approximately 3 inches from stock.

For this job, we have a hoist/lift to raise the car and perform our work comfortably. If you don't have access to a lift, you need to adjust the process to suit your specific needs.

You may choose to remove the entire rear axle housing. You can do this by supporting the rear axle housing with an extra jack stand. If you're replacing one spring at a time you can leave the parking brake cables and rear body brake hose attached, but be careful not to pull the hose too tight. Always remove the driveshaft because if the rear moves too much, the yoke could unexpectedly pop out of the transmission and make a huge mess.

Place a floor jack or some other strong support under the center of the fuel tank before removing the straps. There is a 9/16-inch nut holding each fuel tank strap in the car between the tail panel and the fuel tank. You need an extremely deep socket to loosen them. Drop the straps out of the way and carefully lower the fuel tank, making sure not to tip the tank toward the rear of the car because the excess fuel will pour all over the place.

With the tank removed, pour excess fuel into an approved fuel container. If the fuel is really old and has

CHAPTER 10

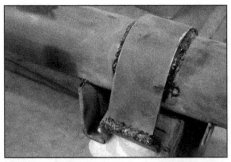

Drag-racing cars back in the day had a steel plate welded from the housing to the leaf-spring perch in order to keep the rear housing tube from breaking off the perch, which was a problem especially on a car running slicks. This differential was found in a 1968 modified by Dana Chevrolet.

turned bad, properly dispose of it. Call your local mechanic or city disposal yard to find out if they allow you to put it in their hazardous-waste recycling containers.

Inspect the top of the fuel tank for remnants of the original broadcast sheet. Also inspect the fuel sender and lines coming out of the sender. If there are fuel stains on top of the tank, you may have a leaking sender O-ring, or the fuel hardline coming out of the sender has broken loose from the sender, which is common.

Common symptoms of these two possible leaks are also the strong smell of fuel and a wet tank when it's full. The fuel filler coupler hose gets very dry and is not reusable after 40 years, even in a well-cared-for car. If it's old and dry, it's a good idea to replace it now, or you'll be doing it again right away. Get the correct kind of hose for this. A section of radiator hose will not stand up to the fuel. Restoration companies sell correct GM-stamped hoses or you can get the non-restoration-correct hose without the stamp.

Rear Sway Bar

In old reference material there was some mention of Chevrolet possibly offering a late 1969 factory rear sway bar, but it never seemed to appear. A heavy-duty Chevy rear sway bar upgrade was available from dealerships. It was offered with the 1967 model-year Camaro, which is obvious because the left sway bar end link bracket was located directly behind the axle over the centerline of the leaf spring. Starting on the 1968, this location is taken up by the left shock absorber. Therefore, if you were to install this heavy-duty sway bar on a 1968 or 1969 Camaro you would have to relocate the left rear shock absorber to the front of the axle and change its mounting points. The instruction sheet included in the kit also shows a multi-leaf rear spring, which wasn't available until 1968. Weird.

Rear sway bars on first-generation F-bodies have been known to induce oversteer. With Chevrolet secretly supporting Camaro racing in the late 1960s, the racing drivers and engineers must not have felt as if a rear sway bar helped much, or they were too busy working on other items that were keeping them from winning every race. With modern-day tire and suspension technology, aftermarket rear sway bars are becoming more popular on Camaros. Hotchkis has designed a complete kit to install a rear sway bar on first-generation Camaros. It requires a couple of holes to be drilled, so for you concours restoration enthusiasts, you may want to stick with no rear sway bar, especially since you won't have the good tires to utilize it. Other enthusiasts can do whatever their hearts desire.

Leaf Spring Installation

1 *Start by safely placing the car on jack stands, so that they don't interfere with removing the rear leaf springs. We had access to a hoist and lifted the body by the rocker panels, which are stronger on the ends than in the middle. To keep from damaging the pinch welds, we placed 2x4 wooden blocks under the rockers. We should have placed some microfiber cloths on the wood to protect the body. (Photo courtesy Mary Pozzi)*

REAR SUSPENSION

Leaf Spring Installation CONTINUED

2 Pull the fuel tank out of the car before removing the leaf springs. Hopefully you don't have more than 1/8 tank of fuel. The more fuel your tank has in it, the more mess you'll have and the more awkward the tank is to remove. On 1967 and 1968 Camaros, loosen the clamp on the filler neck hose that is accessible through the hole behind the license plate. The 1969 neck is soldered on and the tank has to be dropped from the front first. Remove the rubber fuel line between the fuel sender tube and the hardline that runs from front to rear. Remove the ground wire screwed to the front of the passenger-side fuel tank strap brace. Disconnect the fuel-gauge sender-wire in the trunk on the trunk floor just below the center of the tail panel. Push it and the grommet through the trunk floor. If you have the capability of pumping or siphoning the fuel from the tank at this point, you should do so before continuing. (Photo courtesy Mary Pozzi)

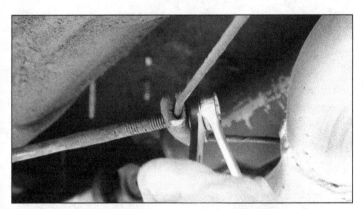

3 To disconnect the parking-brake cable, remove the adjuster clamp on the driver's side of the inside of the subframe. The large round ends of the cables pull through the access hole in the parking-brake cable clamps. Remove the clamps, adjuster hardware, and cable tension rod and put them in a bag for safe keeping. Inspect the cable to make sure it's in good working shape and not frayed badly.

4 Before removing the driveshaft, be aware that it weighs about 15 pounds, and at least a quart of fluid will pour out of the transmission tailshaft when you pull the yoke out. Therefore, you need a clean drain pan. If you have a spare yoke or special transmission plug, you should stick it in the tailshaft to keep out debris and keep fluid in the transmission.

5 Having a helper is a benefit for this procedure. Have him or her hold the driveshaft while you work on unbolting the straps that hold the universal joint caps to the rear pinion yoke. Be careful to not allow the joint caps to fall off the universal joint. Use a long strip of masking tape to wrap the caps in place. If a cap falls on the floor, all the needle bearings may fly out and get lost, or at least get foreign debris in the bearings.

Leaf Spring Installation CONTINUED

6 Slowly pull the driveshaft yoke out of the back of the transmission. After the shaft is removed, check the U-joints for excessive wear that warrants replacement. Wiggle the caps back and forth a little. If there's a discernable amount of side play, replace the joint.

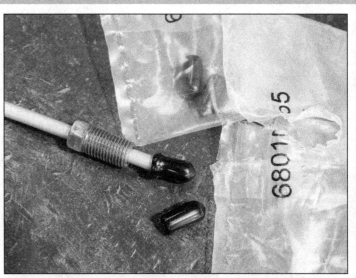

7 Before you disconnect any brakeline you should realize that brake fluid attracts moisture, which is extremely bad for brake components. Closing off the brakelines with caps and plugs helps protect your system components during the project, especially if it's a long-term one. Disconnect the rear flexible brakeline and plug the end of the hose and the end of the brakeline.

8 Remove the nuts from the upper shock-absorber stud located inside the trunk, and remove the lower shock bolts. Check the shock for excessive wear. If there's fluid on the outside of the shock or on the shaft or you can easily compress the shock by hand, consider replacing it. Always replace worn-out shocks as a pair.

9 Remove the lower shock plates and radius rod if you're working on a 1967 that has them. The previous owner had installed lowering blocks and put extra-long U-bolts holding the assembly to the leaf springs. Tall lowering blocks are commonly used, but they adversely affect the handling of the car. We're installing Hotchkis rear leaf springs that are designed to lower the car without the use of lowering blocks. Stock-height leaf springs sit much higher than most enthusiasts prefer. (Photo courtesy Mary Pozzi)

REAR SUSPENSION

10 We decided to support the differential and replace one leaf spring at a time. If you're working without a lift and choose to remove the entire rear differential assembly for additional work, you can balance it on a floor jack and slide the assembly out the side over the top of the leaf springs and set it aside for later inspection. (Photo courtesy Mary Pozzi)

11 Spray some penetrating oil on the three bolts holding the front spring-eye bracket to the floorpan, then remove the bolts. If they don't come out, follow the instructions in the text. Be sure to support the spring so it doesn't fall when you remove the front bolts. (Photo courtesy Mary Pozzi)

12 Support the spring and remove the nuts from the leaf-spring shackles. We chose not to remove the fuel tank to remove the rear-spring shackle hardware, but we had to remove the tailpipes. Unlike typical tailpipes, this custom exhaust has some quick-disconnect clamps on the rear of the mufflers. (Photo courtesy Mary Pozzi)

13 If you're keeping the spring or want to remove the rubber leaf-spring bushings on your own for installing polyurethane bushings, there are two ways to remove them without a hydraulic press. The unsafe and non-environmentally-friendly method is to use a torch to burn them out. The safer and more time-consuming way is to drill them out with a 3/8-inch drill bit. Drill the rubber a little at a time and at different angles. Use a hammer and chisel to remove the metal sleeve that's left in the spring eye.

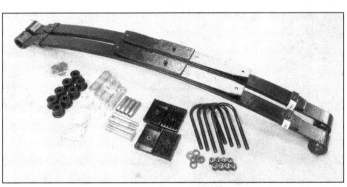

The Hotchkis rear suspension kit came with leaf springs that are 1½ inches lower than stock (also available 3 inches lower than stock), bushings, heavy-duty shackles, J-nut clips, U-bolts, and grade-8 hardware. The lower plates on the Hotchkis rear leaf springs are longer and help keep the springs from wrapping up and increases performance drivability. Not shown is the rear sway bar and HPS 1000 shocks that have been built to Hotchkis' specs.

HOW TO RESTORE YOUR CAMARO 1967–1969

CHAPTER 10

Leaf Spring Installation CONTINUED

14 Paint and refinish all the necessary parts and underbody panels to your liking. The factory mono-leaf springs were painted light gray and the multi-leaf springs were dark gray from leftover oil used to cool the steel after the annealing process. The rear axle assembly, front spring-eye brackets, parking-brake cable brackets, and lower shock plates were semi-gloss black. The T-bolts and U-bolts were black oxide and the lower shock-plate nuts were zinc-plated. The stock shocks were originally semi-gloss gray and the lower attaching hardware was gray phosphate.

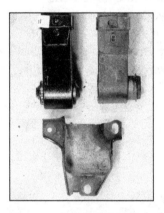

The front bushing had shifted out of the spring that came out of the car (right). The Hotchkis spring came with a rubber bushing already pressed into the front spring eye and the kit came with polyurethane bushings for the rear eye. This is a great combination; the majority of rear suspension shock is transferred through the front bushings and because they are rubber the shock is diminished. The polyurethane bushing in the rear is less flexible and keeps the spring from twisting and buckling under hard cornering. (Photo courtesy Mary Pozzi)

15 If we were using rubber bushings in the leaf springs we would have them pressed in at a shop, but since we are using polyurethane bushings, we can install them ourselves without a press. Install the bushings with the supplied lubricant. The lube is messy, so lube only the parts needed as installation progresses. Push the bushings into place and tap them with a mallet if necessary to seat them fully. (Photo courtesy Mary Pozzi)

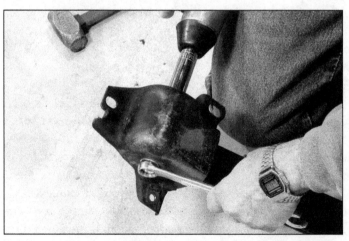

16 Install the front spring-eye bracket on the leaf spring. Since we're installing polyurethane bushings we added the supplied lubricant to the bushing sides because they ride against the metal bracket. Torque the nut to 70 ft-lbs now, because you won't be able to get a torque wrench on the nut once the spring is installed. (Photo courtesy Mary Pozzi)

17 Hold the leaf spring in place and assemble the rear shackle. The nut for the top shackle bolt goes on the inside (fuel tank side) and the lower nut is on the outside of the shackle because of clearance problems between the shackle and the fuel tank. Install the nuts but don't torque them yet. (Photo courtesy Mary Pozzi)

REAR SUSPENSION

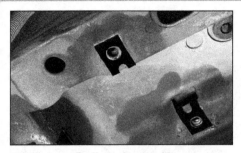

18 Install all six J-nuts. You can re-use any of your original nuts if you choose, but we highly recommend installing new ones. Our particular Camaro is not going to be judged at a show so we installed all new hardware, which was included with our Hotchkis kit. Install the front of the leaf spring and torque the bolts to 25 ft-lbs. (Photo courtesy Mary Pozzi)

19 Maneuver the rear axle housing into the proper position over the leaf springs. Install the isolator pads on top of the leaf springs and lower the axle over them. Install the lower isolator pads and shock plates. Make sure the pads are fully seated on the leaf spring and over the centering-pin hardware.

20 Now install the T-bolts or U-bolts and torque them to 40 ft-lbs. If the lower shock plate is not touching the spring perch and you're nearing full torque, refer to leaf-spring isolators on page 126. Don't bend the shock plate trying to reach full torque.

21 Install the upper shock-mount flange to the body first to 10 ft-lbs. The factory added a bead of weatherproof sealer around the flange and sealing washers around the bolts to seal moisture out of the trunk. Install the shock to the flange. Then install the lower shock bolt and tighten all the top shock hardware to between 5 and 10 ft lbs. The shock isolators should only bulge a little, not squish completely.

22 Lower the car to the ground to allow the suspension to completely compress. Push the rear of the car down a few times to get it to settle. Torque the shackle bolts to 50 ft-lbs. (Photo courtesy Mary Pozzi)

23 If you don't have a good way to support the tank like this, use a floor jack and a wood block to distribute the weight while lowering and installing the tank. David Pozzi lowers the tank with a custom-made jacking plate. When re-installing the tank, be sure the sender wire doesn't get pinched between the tank and body. We taped our wire to the top of the tank to keep it in place. Tighten strap nuts to 80 to 110 inch-pounds. (Courtesy Mary Pozzi)

Leaf Spring Installation CONTINUED

24 We replaced the tank, sender, straps, tank isolator pads, and filler-neck hose coupler with new parts from Year One Inc. Before installing the tank, push the neck-hose coupler up on the filler and tape two clamps on it. This keeps them in place until after the tank is in place. This method only works on the 1967s and 1968s. (Courtesy Mary Pozzi)

25 Feed the sender wire through the trunk floor and connect it, then press the grommet into place. Drop the filler hose coupler from the filler neck and push it onto the tank about 3/4 inch. Remove the tape to drop the hose clamp and install it. Attach the ground wire to the strap brace on the front of the tank and install the fuel hose from the sender to the body hardline.

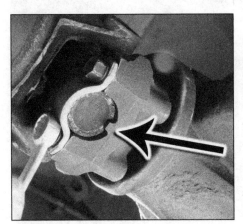

26 Install the driveshaft in the tailshaft of the transmission while keeping the universal joint caps from falling off the joint. Install them in the rear pinion yoke saddles. When you install the straps, make sure the U-joint cups are fully seated between the little tabs in the saddle. Torque the straps to 16 ft-lbs. Install the radius rod if you have a 1968 equipped with one.

27 Install the wheels and tires and all the rest of the parts you may have removed for other work. After you double check to make sure everything is torqued, you're ready to test your new suspension. Double check the torque on the lug nuts after driving 50 miles. (Photo courtesy Mary Pozzi)

CHAPTER 11

ELECTRICAL

The simple word "wiring" often strikes fear in the hearts of many first-time restorers because the idea that the large ensemble of colored wires, clips, and plug-ins seems too complex and can cause even the most daring to seek help. But, if you're unfamiliar with electrical principles and procedures, it simply takes some time to learn the fundamentals and patience to work through the various projects. CarTech's *Automotive Wiring and Electrical Systems* by Tony Candela provides complete instructions for working on electrical systems. There are so many electrical components aboard a Camaro that we cannot offer a comprehensive restoration procedure for each component. However, we highlight important areas of the electrical system, so your particular system is safe, reliable, and appears much like stock.

Safety Tips

When mechanics work on late-model vehicles they are always instructed to first disconnect the battery before turning a wrench. They do this to prevent damage to the ECM (electronic control module) and other electrical items that are very sensitive to incorrect voltages. They are also instructed to do this to avoid causing a fire in case there was a fuel leak; the last thing you want is to have a spark produced by an accidental short. Whenever you're turning a wrench on your older Camaro you should follow the same rule.

If you're working around a serviceable lead-acid battery, always wear eye protection when disconnecting it. In the United States, there are more than 2,000 injuries each year from lead-acid batteries. About half of all those injured get acid thrown in their face and eyes, and this can happen when a battery explodes. Lead-acid batteries produce extremely flammable hydrogen and oxygen gasses when they are charging, due to electrolysis. A spark caused by connecting or disconnecting a battery terminal when these gasses are present will ignite them. The spark you see when disconnecting a battery is caused by a current draw being present from an

During the last days of May in 1969, Chevrolet switched from top-post batteries to side-post batteries. This reproduction battery has the correct Delco Energizer logo and service caps. The only drawback is the old technology inside isn't great for extended periods of non-use.

HOW TO RESTORE YOUR CAMARO 1967–1969 135

CHAPTER 11

accessory being on. When you think everything is turned off, don't forget that a clock, stereo, dome light, etc., also create a draw on the system.

When disconnecting a battery, always disconnect the negative battery cable first; when connecting a battery, always connect the negative battery cable last. This keeps you from accidentally shorting out the electrical system by touching the wrench to ground when you're disconnecting or connecting the battery. The battery has an incredible amount of energy and touching your wrench directly to the surrounding sheetmetal causes a lot of sparks and turns your wrench into a red-hot skin-burning implement. The same goes for the wrench when you're disconnecting the power wire on the alternator. Other smaller-gauge wires do not produce the same amount of heat; they typically just pop the fuse, if there's one present.

Another important safety tip is to always remove your jewelry—rings, watch, and other metal items on your wrists and hands. These quickly become conductive items if they get caught between a piece of metal and a positive terminal on the battery, alternator, or hot junction box. People have had severe burns from jewelry that got welded to metal or turned red hot in a split second from connecting hot terminals directly to ground.

In a nutshell, here are the four rules of electrical safety:

- Wear safety glasses when disconnecting a battery.
- Disconnect the negative battery terminal first.
- Reconnect the negative battery terminal last.
- Don't wear jewelry when working on a car.

Battery

All first-generation Camaros came from the factory with a lead-acid battery, which are still used in cars today. Battery design has changed over the years. Original first-generation Camaros with serviceable batteries had caps on the top so you could add water to the cells. Current batteries are non-serviceable; they actually last longer than the old ones. Old batteries were more prone to have acid make its way out of the service caps and cause additional corrosion, which is why most original battery trays are eaten away. Newer batteries can still cause problems, but do so less often than the older ones.

From 1967 through the last part of May of 1969 (by build date) the Camaro used top-post batteries. By June of 1969 Camaros were completely switched over to side-post batteries. It seems insignificant, but when you're restoring a car to be factory correct, it's something you should take it into consideration when restoring a basket case that might have already been stripped of its battery cables.

Reproduction

If you do care that the battery looks stock, companies like Year One and Classic Industries sell reproduction batteries with the proper screw-on service caps and labels.

Aftermarket

If you're a hobbyist who doesn't care about everything looking factory, you should consider installing an Optima Batteries Spiralcell battery. The old technology of lead-acid batteries has been eating up battery terminals, battery trays, and inner front fender wells of Camaros since 1967. The Optima battery is sealed and is made of spiral-wound Absorbed Glass Mat (AGM) technology. The popular batteries are: Red Top (which is for everyday use as a high-output starting battery), Yellow Top (more for high-draw systems and for cars that may not be driven a lot because it's less susceptible to failure after discharging), and Blue Top (for marine applications). If you will not be driving your car every day, you want to consider the Yellow Top for the best performance. If you have additional high-draw accessories like aftermarket electric fans or a powerful stereo system, the Yellow Top is the best battery for you.

If you don't care that the battery looks original and aren't going to drive your car on a daily basis, but want an extremely reliable battery, pick up an Optima Batteries Yellow Top. It's a deep-cycle battery, so if it drains, you can recharge it back to full energy capacity time after time.

Installing the Harness

Camaro wiring usually consists of five or more harnesses that all plug into one another at some point. If your car has additional options you can expect additional wiring harnesses. Think of the wiring as the nerve center of your Camaro. Without the wiring, nothing will work on your car. But if you pay close attention to routing, and

ELECTRICAL

have your trusty assembly manual (and this book) by your side, you can effectively diagnose problems and come up with solutions.

If you are reusing your factory-original harnesses, please look them over closely. Many harnesses have been modified for accessories over the years. This does not mean that your wiring is unsafe, but pay close attention to poorly wrapped wires, splices, melted wires, missing insulation, broken plugs, or cuts. Animals like mice can cause damage from chewing on your harness. If you have too many splices or other damage, a new harness is often the best option.

Remember that a damaged harness can cause a fire due to excessive resistance caused by a bad connection or a bare wire that touches a ground. These issues can ruin the best restoration. If your original harness is severely damaged, and cannot be repaired, save it because you may need to use some of your old connectors and clips for ones that are missing in an aftermarket replacement harness.

Under-Dash Wiring

The first harness to install, and possibly the most important, is your under-dash harness, which includes your fuse box. Pay close attention to the fuse box because it safeguards the wiring.

You need to be certain that the box is solidly mounted. It typically mounts through the firewall and has two locating tabs on either corner. When installing this part, make sure it is not hung up on the firewall pad. Also, make sure the holes for the special screws that mount the box are not oversized or stripped out. If one or both of these are oversized, you will not be able to securely mount the box. Once you have determined that the holes are in good shape, mount the box.

Pay close attention to the fuse panel; in particular, what amp fuse is used in each specific location. The panel is marked for the recommended size. Do not substitute larger-amp fuses; this commonly causes fires. I have seen everything from coins to gum wrappers substituted for fuses—not a good idea. The fuse is the physical safeguard of the electrical system. Each operation has a rated amperage. If you exceed the amperage with a larger fuse, it can cause a short, and then the fuse will blow. If you have a fuse that blows over and over again, check for a grounded-out wire or a bared wire. This is often the cause.

To install the dash harness, you can lie under the dash and reinstall each connection, or you can pre-wire the dash on the workbench and install the entire harness at once. Either way requires about the same amount of work.

GM did a great thing by making sure the wiring is plugged into each switch correctly. GM designed each plug to only match up to the right accessory. Each plug has its own unique shape, and cannot be installed into the wrong switch. If you are having trouble plugging something in, one of two things is wrong: Either you are forcing the wrong end into the switch, or you have a bent tang. If you are forcing the connection, recheck your work. Sometimes you can get the connection made, but the tang has bent, thus not making contact. Because this mistake is not easily seen, it is often difficult to trace.

Trunk Harness

On 1967 and 1968 models the trunk harness includes the intermediate harness. This is a separate part on the 1969 model. The trunk wiring is fairly straightforward.

At this time, you must also run the license plate lamp wire and the gas tank sending unit wire. Both enter through the trunk floor or the rear bumper panel. If you have an RS-equipped car, the reverse lamps wiring comes up through the trunk pan near the bumper bracket braces.

Pay close attention to what bulbs you use for the taillights. Using the incorrect bulb can create too much heat and melt the taillight lens. (The same goes for every light in the car.)

If you installed a new headliner, make certain you also ran the dome lamp wire. If you forgot to install that wire, it would be difficult to install now, unless you take the headliner down.

Engine and Headlamp Harnesses

The engine and headlamp harnesses plug into the main fuse box at the firewall. Once again, be very careful that you have a smooth fit when you plug the two halves into the main box. If you need to force the connection, look closely at the connections. Make sure you are not bending anything. Once you establish a good connection, go ahead and tighten the center bolt snug, but do not overdo it.

Refer to your assembly manual for the correct routing. Headlight harnesses run behind the inner fenders on 1967s and 1968s; and in front on 1969s. Every so often you will see a clip on the harness. The clips fit into pre-punched holes along the core support and fenders.

The engine harness runs along the firewall and includes the wiper motor, coil, distributor, blower

CHAPTER 11

Important Tips

Always pay close attention to the grounds because electrical circuits short out without a solid ground. Because most restorations have fresh paint and built-up layers of primers and other materials, you need to verify that there is a clean ground if you are having trouble with a connection. A test light is a good way to test your work, if you have trouble. More often than not, the failure is the result of a bad ground.

Throughout the harness there are ground wires. You have to make sure you have a good bare surface to achieve a proper ground. Use the special ground washers where provided. Electrical washers dig into the metal to provide a ground. The more options your car has, such as air conditioning, ZL/2 hood, console with gauges, etc., the more wiring you have. Just pay attention to the proper installation procedures for achieving a solid ground and you will get a great result.

If your goal is to make your harness as correct as possible, and you have bought a new one from the many vendors out there, and you think it's correct, think again. Check out these differences:
- Lay out your original harness next to the reproduction. Notice anything? Many of the reproduction harnesses do not use the correct ends and clips, in both color and configuration.
- Look at the original headlamp connectors and you will find they are brown, the repos are clear.
- See the locating clips? Most often they are white, while the originals are brown and black.
- Look at the plug-ins for the wiper motor; the spacing is just a bit different.
- The same holds true for the trunk connectors.

Always compare your originals to the aftermarket replacements. In some instances, reproduction harnesses are completely missing certain clips, so you're required to install some of the connectors and clips from your original harness. This is one reason you should not throw away your old harnesses until the entire job is done.

If you want to use your old parts, it is not that hard. Unplug the old ends and install them onto the fresh harness. Clean them and inspect them for cracks. You may have to modify a small screwdriver to unhook the wires.

While you are at it, do not overlook the harness tags. GM installed small part-number tags and codes on the wires to help identify the harness. If possible, reuse your old tags. It takes just a bit more time, but in the end it may be worth it to you.

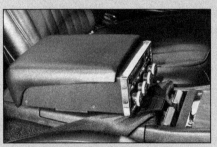

The Camaro was available with an 8-track player that could be mounted to the bottom of the dash or on top of the console. The pad on top of the unit was only available in white, so this one has been painted to look better, but it's not correct.

The factory wiring harnesses used black (arrow) or brown anchors to hold wiring in place. Reproduction wire harnesses frequently have unsightly and incorrect white anchors. Many ground wires require the star washer against the panel to bite through the paint and create a useable ground connection.

motor, and starter. Small metal clips that are rubber-dipped secure the harness to areas like the heater box, etc. Use them where necessary. This will keep the harness away from high-heat areas like the exhaust manifolds. Extreme heat easily melts the covering and causes the wire to fail.

Ignition

You can use original factory breaker points and worry about gap and dwell. Or, you can step out of the dark ages and upgrade to a high-performance no-maintenance electronic ignition system with little or no visible difference.

Pertronix Ignition

You can swap the antiquated breaker-point system in your stock distributor for an electronic ignition. Pertronix Performance Products offers the extremely popular and performance-proven Ignitor points-to-electronic-ignition conversion kits for early Camaros. One additional

ELECTRICAL

This original distributor was upgraded with a Breakerless.com stealth electronic ignition that only needs a single wire to power it.

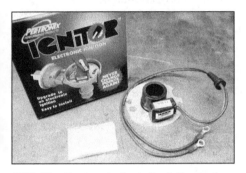

With the Pertronix installed, the only telltale item on the outside of the distributor that it's not original is the two wires coming out of the bottom (instead of one wire).

wire coming out of the bottom of the distributor is the only physical evidence that it's not stock. The stock points have a single wire sticking out of the bottom and the Pertronix ignition conversion kit has two wires, a positive and a ground wire.

Performance of stock points begins to taper off at about 5,500 rpm for a decent set, and even lower for an older set. If you've ever had to mess with points, you know that they can be finicky and need to be replaced every time you perform a tune-up. With an electronic ignition you never have to mess with points again. You still need to service the cap, rotor, and wires, but the module inside the distributor is a one-time replacement.

Ignition Coils

The original coils have raised letters and numbers stamped in the case and the Delco-Remy product name on the top. Ignition coils that are close to stock in appearance yet provide substantially stronger spark and, thus, performance are readily available for the first-generation Camaro. Of course, Accel, MSD, and others offer ignition coils that provide substantially higher outputs. The Super Stock works with the original breaker-points ignition system as well as the new CDI systems. The high-quality windings and design provide even higher energy output, plus improves throttle response and top-end power.

Alternator

The standard original alternator was only rated to output 3 amps. Some alternators have a stock-like appearance and provide substantially improved performance. Several companies, such as Tuff Stuff, offer high-amp alternators, which are ideal for a high-amp MSD, Accel, or similar ignition boxes. Of course, if you run a premium stereo system, the alternator provides enough power for these power-intensive systems. The heavy-duty Tuff Stuff is a 140-amp unit that features jumbo diodes with bi-directional cooling fins, a spin-balanced heavy-duty copper rotor, an effective cooling fan for diode cooling, heavy-duty bearings, and a bar stock single-groove pulley.

Tools

The only electrical tester tool you'll ever need is the Power Probe III. It connects to the battery with two large spring-loaded clamps and has a 20-foot cord, so you can connect it to the battery in the front and test circuits on the other end of the vehicle. Also, accessories don't even have to be attached to the vehicle, so if you want to test a wiper motor on your bench, you can.

The beauty of the Power Probe III is that it has a rocker switch that changes the pointed testing tip into a 12-volt source, a ground, or a normal testing mode. This makes it really easy to add power or a ground to an accessory or a circuit in order to test it by turning it on and off. The tool has a built-in quick-reset breaker so if you accidentally touch a hot wire, you won't damage the tool or your wire harness.

If you're working in the dark, like under a dashboard, this tool has two bright LED lights to use as a flashlight. There's also a built-in screen that shows how much voltage is in the circuit you're testing. For continuity testing, it has a separate lead you attach to a circuit and then use the tip of the tester.

This Power Probe III electrical tester replaced my test light, continuity tester, and volt meter. It's got a digital readout, resettable circuit breaker, bad ground indicator, and a switch to power the tip with 12 volts or to make it a ground lead. The 20-foot test leads allow testing on the opposite end of the car or on a workbench. It's invaluable for wiring cars.

HOW TO RESTORE YOUR CAMARO 1967–1969

CHAPTER 11

The brown headlight connector on the left is the factory connector and the clear plastic one on the right is a common one used by wiring reproduction companies. If you still have your original harness, you can swap the connectors.

Most terminals like this one are locked into the connector by a small tab that depresses in order to release it. This little tang on the connector keeps the terminal in the headlight connector so it doesn't pop out when connecting it to the back of the headlight. Be sure the tang is sticking up slightly before placing it in the new plastic connector. Don't bend the tang too much or it will break off.

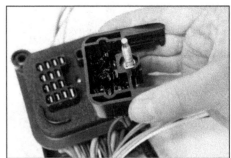

Make sure all the contact pins are straight and have a good contact surface before plugging the harnesses together. Never force them if you feel resistance during assembly. If a pin moves or is out of place you could damage the terminals. If there's resistance you can wiggle the connector a little in an attempt to line up the pins. The black grease used here doesn't conduct electricity, but helps keep moisture out of the connection.

Like many project Camaros, ours sat outside for years. The original fuse box has a lot of corrosion, so this new AAW Factory Fit fuse box and wiring harness were installed. Many original harnesses have been butchered and overloaded by shade-tree mechanics and might have faulty wires, so check to see if a replacement is in order.

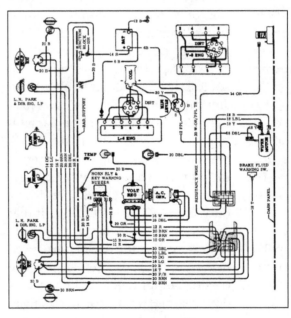

If you're tracing wires to find an electrical problem, do yourself a favor and purchase a Chevrolet Chassis Service Manual. It has all the wiring diagrams broken down by component, switch, pin location in connectors, wire color, and wire gauge. It drastically cuts repair time. (This image is reprinted with the express consent of Year One, Inc, a licensee of General Motors Service Operations)

American Auto Wire Factory Fit makes all the wiring harnesses necessary to restore your car.

Other testers have an LED light that switches from red to green when you transition from voltage to a ground. If you're color-blind, this makes testing circuits very tough. The Power Probe III takes care of that problem with separate lights so you know exactly what you're testing.

Power Probe also sells tools for other specific tasks, such finding shorted or open circuits with their ECT2000. For more info, visit www.powerprobe.com.

140   HOW TO RESTORE YOUR CAMARO 1967-1969

Non-Factory Wiring

We can't stick our head in the sand and assume everyone will stay with their stock wiring harness or that they won't add electrical components. Other than the small percentage of concours-quality show cars, many regularly driven Camaros have an aftermarket stereo, electric fan, gauges, ignition, or electric fuel pump. The following are details and topics that are important for sound operation.

Relays

Electric fans are probably the most common accessory wired incorrectly. Common electric fans come in two types: large-diameter high-CFM fans and low-CFM fans. The high-CFM fans draw a high-amp load and effectively cool an engine; small-diameter fans flow low CFM and are very ineffective at cooling more than a 6-cylinder engine. Good fans draw more amps and should be wired with a relay.

Running an 18-gauge power wire from a switch on the dash or even one from a couple of feet away from the battery will overload a small-gauge wire. A small wire over a long distance, powering a high-amp-draw accessory, overheats the wire and creates more resistance and more heat. This drops the amount of power actually reaching the accessory and also could create enough heat to melt the insulation off the wire, short against other wires or metal, and cause an electrical fire.

You need to have a short distance on high-amp-power accessory wires. The best way to do this is to install a relay to power your high-amp-draw accessory. You should also install a relay for low-amp draws too. It's just safe practice.

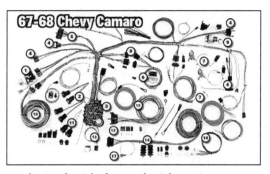

American Auto Wire makes custom wiring kits for the first-generation Camaro (and most any popular car) that replace the original fuse box with one that uses late-model blade-type ATO fuses. These kits also include additional circuits so you can easily wire in other accessories such as electric fans, electric water pump, custom gauges, etc. All the wires are labeled with words printed directly on the color-coded wires.

Relays allow you to use low-current switches to operate high-amp accessories for reliability; they come in a few different configurations. This diagram uses a typical Bosch 0332019150 with two 87 terminals (the center 87 isn't labeled in the diagram) that power up when the switch is turned on. The additional 87 can be used to power another accessory. Always protect a circuit with a fuse.

Installing an accessory with a relay allows you to have a low-amp-power wire from a switch (usually located many feet away in the interior of the car, such as the dash or console) that powers the relay.

Electrical Components

The best source I've ever found that has just about every electrical component for wiring an automobile is Del City (www.delcity.net). It sells relays, fuse boxes, cross-linked wire, WeatherPak connectors, switches, tools, and complete wiring accessory kits. Its website has all the important details about amp ratings and physical dimensions of its components to help you find exactly what you need.

Wire

The wire you buy in the little packages at big chain auto parts stores is typically cheap wire. It's not necessarily the wire that's cheap, it's the insulation over the wire that's cheap and low quality. It gets the job done in most cases, but cross-linked wire is much higher quality. It stands up to dirt, oil, and temperatures up to 275 degrees F. It resists kinking, which is especially helpful when pulling it through panels. A kink in a wire can damage it by creating resistance in the wire, which is especially bad if you have finicky high-tech systems, such as fuel injection.

The best aftermarket wiring harness companies use cross-linked wire to build their harnesses. Late-model cars use cross-linked wire exclusively because of its quality.

Fuses

If you're adding an electrical accessory you're going to, at the very least, install a power source and a ground wire. (The only exception to this rule is when you're splicing into a circuit that has an existing fuse.) You should never install a positive wire without a fuse between the voltage source and the accessory. The fuse is protection against electrical fires that can be caused by a faulty component or wiring, a collision that pinches a wire, or simply grounding a wire with a tool when working around an exposed terminal.

There are a few different types of fuse holders. If the holder is located outside of the interior, it's a good idea to use a weatherproof fuse holder.

Custom Wiring

Wiring isn't as difficult to understand as most people make it out to be. You just need to educate yourself. *Automotive Wiring and Electrical Systems* by Tony Candela provides an easy-to-understand explanation of automotive electrical systems, as well as detailed diagnostic descriptions and solutions for many common electrical problems.

As you become experienced with wiring and understand more aspects of the whole system, you might build your own custom wiring harness using some existing factory wires and adding your own by safely splicing into original wires or adding your own fuse box. If you're going to attempt this, you need to examine the wiring diagrams for your car. Take a little guidance from them and draw up your own diagram for every custom wire you add. This helps when troubleshooting your harness during assembly and, later, if an issue pops up or you want to add another component.

Aftermarket Accessories

The 1967 RS headlights have electrically operated headlight doors, which was a great idea but had its shortcomings. Chevrolet switched to vacuum-operated headlight doors for 1968 and 1969 Camaros.

The vacuum system was bulky with the vacuum reservoir tank, actuator cans, valves, and hoses. All these parts must be operating correctly in order to have the headlight doors open when the lights are turned on. In most cases, owners just give up on the vacuum system and open the headlight doors by hand when night falls.

Fortunately, Detroit Speed offers a kit to convert the 1968 and 1969 cars to electric operation. It utilizes modern technology to sense when the doors are fully open or closed and cuts the power to the actuators. This is different from the original electric operation in the factory 1967s, which had limit switches at full-open and full-closed that periodically failed. That stock limit-switch failure causes electrical problems.

The Detroit Speed kit is much less bulky and complex than the original system. It's made of two small actuators, mounting brackets, hardware, wiring, and the electrical controller. This kit also helps if your headlight doors don't work because you have a big camshaft or low vacuum.

If you plan on driving your car on a daily basis, you will use your windshield wipers once in a while. After many years of driving late-model cars, you can get used to creature comforts like intermittent wipers. Unfortunately, the factory Camaro wiper system leaves a lot to be desired. It has three settings: off, on, and ridiculously high. When you're driving around in light mist or sprinkling rain you have to constantly turn on your wipers every few seconds, because the "on" position is just too fast for light precipitation conditions. This on-off-on procedure gets annoying after a while.

Detroit Speed sells its Selecta-Speed wiper kit for all three first-generation Camaros. It replaces your original wiper motor with a late-model unit that has five delay speeds plus low and high. The kit includes a custom knob, stand-alone wire harness, and wiper motor. Another great part of the design is that it fits flush up against the firewall so you can run larger valve covers on your big-block.

If you drive your Camaro everywhere and in all kinds of weather, you'll appreciate Detroit Speed's Selecta-Speed wiper kit. It has five delay speeds, plus low and high. Therefore, you do not need to make your own delay by constantly turning your wipers on and off in light-precipitation conditions.

CHAPTER 12

INTERIOR

Interior restoration is different from every other aspect of classic-car restoration because it's not mechanical. And as such, many at-home restorers and mechanics are not experienced or knowledgeable with this particular area of restoration. But with some patience, the correct materials, and the right tools, many home restorers can successfully replace door panels, rebuild worn out and torn seats, and replace ratty old carpet, replace a dash, and successfully complete most other projects. Let's face it the interior is where the driver spends his or her time, and the condition of the interior directly affects how we perceive the driving experience and the car as a whole. Therefore, a like new factory interior is not only desirable to maintain collector value, but it projects a positive image of the car. In this chapter, we will cover how to replace, refurbish, or repair major interior components and explain the techniques, tools and materials required for restoration of your Camaro's interior. For reproduction parts, classic Industries, Year One, and others may have the parts. Al Knoch also carries made in the USA Camaro interior parts.

This is a 1969 center console (RPO D55) with gauges (U17). The woodgrain trim is (part of RPO Z87) on the cluster, console, steering wheel, and the passenger side of the dash. (Photo courtesy Tony Lucas Collection)

Headliner

By now your headliner has been replaced or it's on its last leg. The thread that holds the fabric panels together is very delicate. After 40 years, one slight touch will snap that thread in the original headliner in an instant and cause the fabric panel to drop right on your head.

The best scenario for replacing the headliner is with the windshield and rear window out of the car. It can be done with the windows in; it's just easier to install with them out. If the rest of your interior is installed, you should take about 15 minutes to remove the front and rear seats before attempting this job.

Important items needed to perform this task are: headliner, headliner glue, binder clips, scissors, and razor blades.

Note: Before pulling out the headliner, inspect how it's trimmed and cut at the corners and how it's glued to the pinch welds. This gives you a good idea of how it was installed and gives you some insight as to how to go about installing a new headliner.

HOW TO RESTORE YOUR CAMARO 1967–1969   143

CHAPTER 12

Headliner Installation

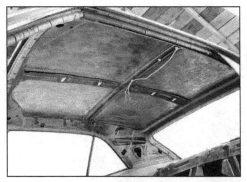

1 Quite a bit of surface rust had started under the roof insulation, which is typical, but we got lucky that it was not pitted or serious yet. We used a molded bristle disc to clean the metal before we applied the KBS coatings to protect it from rusting again.

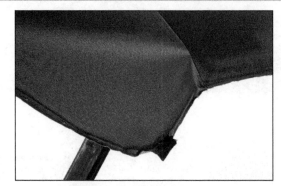

2 Install the headliner in all five bows and clip them in place. If the plastic clips are broken or missing, the headliner does not install correctly and sags. Be sure the headliner is centered on the bows before moving to the next step.

3 Install the three sun-visor screws before installing the headliner. Once the headliner is installed, press the headliner up and cut a small X on the top of the screw to get it out. This is much easier than compressing the headliner to find the small screw holes.

4 Use the same method of pulling and clipping the headliner on the side pinch weld. Once all four edges are pulled tight, check to make sure you're happy with the headliner fitment, with the exception of the corners. Adjust the clips as necessary.

First, remove the A-pillar covers and pull the vinyl trim off the pinch weld around the sides of the headliner. Next, remove the sun visors and rearview mirror. Remove the plastic interior-color trim around the rear window. The sun typically destroys the interior window trim. If you have shoulder harnesses, pull them out.

The sail panel covers are held in place by the pinch-weld vinyl and by clips similar to the ones holding the door panels in place. By sticking a thin prybar between the door panel and the sail panel you should be able to gently pry the clips out of the sail panel. The four clips that hold the panel in place typically break off the back of the sail panel cover (because of the age of the parts). There are two clips on the top and two on the bottom.

Place your prying tool under the square metal anchor that is attached to the sail panel cover. Don't pry the cover out too far because it will snap off. If you snap the cover or break the anchors loose from the cover, you'll have to use completely new panels instead of reupholstering your originals. Note that the new panels are not great in comparison to your originals.

The headliner is held in place by sprung steel bows, which press into plastic clips attached to the roof support on the ceiling. Try not to mix up the five metal bows when you remove the old headliner—they attach in a specific order.

Four of the bows slide into the side of the inner structure of the roof. At each of these locations, there are three holes. If you didn't mark which hole each one came from, the headliner might not install correctly, so it may require some trial and error to properly install it. Don't worry if you mixed up the four pre-bent bows. You can use a tape measure to figure out which bow goes where.

INTERIOR

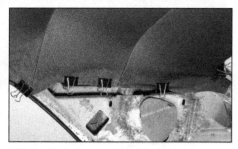

5 Pull the slack out of the material above the sail panel and attach the headliner to the tack-strip in the sail panel by wrapping it around and gluing it or by anchoring it with heavy-duty staples. Perform the same work on the other side and make any adjustments to the clipped headliner.

6 If there's one small spot you just can't get tight no matter how much you adjust the material and clips, you can use a hair dryer to heat the headliner a little to shrink small sections, but only do so after the headliner is completely installed and glued in place.

7 The glue you use for attaching the headliner is contact cement, which means you coat both pieces and stick them together. Brush-on adhesive is preferred over spray-on types because it typically provides a stronger bond. Using the proper tucking tool eases installation.

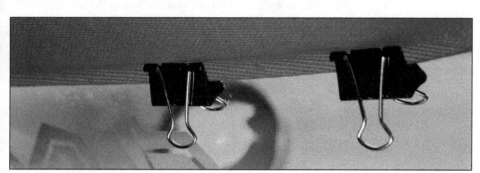

8 Pull the clips out of a few feet of the headliner and apply a film of adhesive to the flange and the headliner material. Once the cement is slightly tacky, but doesn't transfer to your finger, it's dry enough to pull the material to the flange and clip it in place again. Leave the clips in place for several hours.

9 Trim around the corners. The rear window typically requires small pie-shaped cuts to get the curved area to look smooth and not bunched up or creased. Cut as necessary, but not so deep that you can see it with the trim installed. Glue the corners and clip in place.

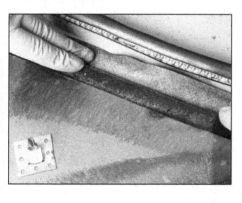

10 We purchased the kit with the material to go over our original sail panel covers. Spray-on contact adhesive goes on a little more evenly for a large surface area. This keeps unwanted lumps from showing up as when applying it with a brush. Leave enough material on the front and rear edges to tuck into the vinyl and plastic pinch-weld coverings.

11 The local upholstery shop was not able to get the stitching correct with the piping and the chrome ends, so we purchased finished visors. With a new dome light and all the edge moldings, A-pillar covers, and accessories installed, the headliner looks like a professional performed this job.

HOW TO RESTORE YOUR CAMARO 1967–1969

CHAPTER 12

Starting from the front:
Bow 1 = 46-1/2 inches
Bow 2 = 46-1/2 inches
Bow 3 = shortest straight bow
Bow 4 = 49-1/4 inches
Bow 5 = 48-1/4 inches

If you have the space, take the headliner out of its box and lay it out flat for a couple of days before installing it. But don't iron it to try to remove wrinkles. Heat shrinks the material. After the headliner is installed, you may need to use the shrinkage to your advantage.

While we had the headliner out, we treated all the surface rust that was behind the headliner on the sail panels and cross-bracing, which snowballed into a much larger job.

We noticed the underside of the roof panel didn't have any paint on it when it left the factory; some of the insulation had fallen out and there was a good amount of surface rust on it. The factory insulation disintegrates the instant you touch it. It's obviously glued to the roof panel before the panel is welded in place because it's between the center braces and the roof panel, which means the underside of the roof panel is bare steel.

Your choice is to leave it alone if it looks decent or make a huge mess removing it. We decided to make a huge mess and scrape it all out. Our project car didn't have any interior in it, which made it much easier for us. If you're thinking of doing this same job and your car has a complete interior, we suggest removing as much of it as possible and covering everything up to the windows with one huge sheet of thick-mil industrial plastic. Protect as much as you can from dust, chemicals, and paint.

Since this job throws a lot of dust and we're not sure what this material is made of, it's best to do it outside. For your safety, use respirators, safety glasses, and long-sleeved clothing.

We scraped the insulation and its glue off the panel with a blunt gasket scraper, making sure not to gouge or damage the steel. After much of the glue was off, we prepped the surface with the same "bristle disc" we used on the floorboards. To keep from producing metal-distorting heat in the panel, we used the pad sparingly and often moved it to different areas.

Clean up all the insulation dust and debris and dispose of it before removing your respirator.

Don't leave the rust lying around because it continues to spread. Take some time to treat the metal with KBS AquaKlean and RustBlast, along with a coat of RustSeal. Once the KBS is completely cured, we installed Thermo-Tec to reduce road noise. (It's the same product we used on the floorboards.) After installation, we ran a small wooden roller over the entire surface to make sure it was stuck to the panel correctly. To give the roof panel extra rigidity, we added dabs of sealant between the panel and the cross-braces.

If you're doing this project with the interior installed, cover everything with a large single thick sheet of plastic before using any KBS products. The plastic allows metal prep chemicals to pool up, rather than dribble onto carpet and upholstery. As recommended earlier, you should perform this job without the seats and carpet installed so you don't risk damaging them, especially if your job turns into treating the rust on the roof.

Place the bows into the correct loops of the headliner, then take the assembly to the car and install the bows are the car. Metal tabs that pierce the looped canvas-type material hold the center bow in place. Make sure the looped material and bow is centered in the ceiling before you pierce the looped material. Bend the tabs to firmly hold the bow in place. Install the rest of the bows in their correct positions on the ceiling side-supports, and place the plastic clips in the center of the ceiling.

Spray-on contact adhesive makes a real mess because there's a good chance of accidentally spraying it on other surfaces. Brush-on adhesive is much cleaner because there's no overspray. The only drawback to applying adhesive by brush is needing to control drips, which happens when you try to use too much of it at once.

With either of these types of contact adhesive, apply a film of adhesive on both surfaces to get a strong bond. Apply the adhesive to the headliner and flange surfaces and allow them to dry enough so that the surfaces are slightly tacky, but doesn't transfer adhesive to your finger with a light touch. When pulling the glued headliner onto the flange, only touch the portion of the headliner that interior trim covers. This way the rest of the headliner doesn't get a bunch of messy fingerprints of adhesive. Clip the headliner in place and leave the clips attached for a few hours afterward to allow the adhesive to dry.

The 1967 Camaro headliner features a material pattern called Impala Leatherette. The 1968 and 1969 cars have a Bedford Ribbed–style finish. Some companies offer slightly cheaper headliner kits if you don't care about the material type. Kits come in two basic types: Some have extra material to cover your old sail panels and sun visors. Other kits have new sail panels, which are not as good as the originals (they don't have attachment anchors) and don't include material for sun visors. You can purchase reproduction sun visors already upholstered to match your new headliner.

Kick Panels

Most kick panels have been abused during decades of driving. As they age and the plastic deteriorates, the Madrid grain starts to scrape off the surface. When the grain scrapes off, a scuff is left behind. If they've been painted a different color, you're left with ugly colored streaks. If your panels were originally one color and painted with vinyl paint, there may be hope to save them if you want the original color back.

Original kick panels are molded in the original factory interior colors. Restoration companies are reproducing these panels, but to keep costs down and inventory low, they only make these panels in black and you have to paint them yourself. If you're restoring a car with a lighter-color interior and can find good original panels, you're better off using them so that if the surface does get scratched, a dark color won't show through.

Original A/C-equipped Camaros have different kick panels than non-A/C cars. There were two non-A/C kick-panel designs in 1969. There was a two-piece kick panel with a separate fresh-air grille until January or February when the one-piece kick panel design showed up.

Interior Paint

Much of the interior is upholstered, but there are also many components that are painted or molded plastic and vinyl. For factory finishes, the interior paint is different than the exterior body paint. If you look at some photos of bodies and interiors, you may think that the interior and exterior are the same color. The only reason is that someone restored it incorrectly or they customized it.

Even if the body color is Bolero Red, and the interior paint is red, it's not the same red—the interior paint is a 30-percent or 60-percent gloss. This means that there are two different levels of gloss paint on the interior. The top of the dash and the metal surrounding the rear package tray have the lowest amount of gloss (30 percent) to combat against blinding light reflecting into the eyes of the vehicle's occupants. The other surfaces in the interior that received paint were done so with a slightly more gloss (60 percent). Factory interior paint was not as glossy or the same color as the exterior main body paint.

Some interior parts were molded in the color of the interior. As of this writing, reproduction companies have not yet started reproducing the harder plastic parts in colors other than black, so if you purchased reproduction parts, you may need to paint them.

The unfortunate thing about painting an interior color over a contrasting color is when the paint is scratched off it becomes extremely obvious, especially on kick panels. Good, original non-black kick panels are not easy to come by, but if you look around a swap meet, you may find some that have been painted with black interior paint. If you're lucky, you can remove the paint with a mild abrasive, such as Fast Orange hand cleaner, and a soft brush so not to damage the underlying finish.

A good tip is to begin removing paint from the surface in a small, inconspicuous spot to be sure you're not damaging the part before you start removing paint in the center. This method works well on plastic that is painted with vinyl spray paint.

Removing vinyl paint from vinyl is much harder because the material is more porous than molded plastic. The consoles, main dash cluster, and steering-column covers are made of harder plastic than the kick panels, rearview mirror trim, seat backs, etc. With the harder plastic panels, you simply scuff the old paint or remove the paint with a light baking-soda blast, primer, and paint with the interior paint of your choice.

If you paint the color-molded parts, make sure to use the correct vinyl or interior paint for fabrics and flexible surfaces.

Door Panels

Other than the seats and the steering wheel, the door panel probably gets the most physical contact in a car. One of the most common areas of door panel damage is the armrest (and surrounding area) because it's the handle used to close the door and a place for resting elbows. The other commonly damaged area is the lower front section of the panel that seems to get a lot of contact from shoes while entering and exiting the car.

Repair

Over the years, there have been multiple ways to fix damaged door panels. You can purchase a vinyl repair kit from Eastwood or other companies, but keep in mind that these kits are only meant to repair small damaged areas and cracks. Results with these repair kits vary because it's not easy to get the right color or to hide the texture of the repair. Another concern is the

CHAPTER 12

amount of use the repair will get; a lot of rubbing and scuffing in the repaired area can damage the repair. We suggest attempting a repair on a test subject or a spot in an inconspicuous location before making a repair in a highly visible area.

If money is no object and you really want to save an original panel, you can send it off to Just Dashes in Van Nuys, California. Just Dashes makes the damage disappear by stripping your original dash and completely rebuilding it to look and feel just like new. It offers this service for most every interior panel.

In the past, flexible vinyl skins were glued over old door panels to hide damage on deluxe door panels. There are also companies that offer hard plastic panels that simply go over your original panel. These hard panels can fool average observers until they get a close visual inspection; and brushing up against a hard panel is a dead giveaway.

Standard vs. Deluxe Panels

If you want to swap standard door panels for deluxe interior panels or vice versa, it requires more than a panel swap. The front door window regulators for the standard and deluxe interior panels are different on all three years. The front door lock pull remote (mechanism where the interior door release handle mounts) is different for the standard and deluxe interior panels on 1968 and 1969 models. The standard release handle on the 1968 and 1969 models pulls upward, and the deluxe handle pulls outward from the panel. You can purchase reproduction door remote mechanisms for both standard and deluxe panels. You can also look for used originals at businesses that specialize in new and used Camaro parts, such as Steve's Camaros.

The 1967 Camaro came with short door panels. The top portion of the interior side of the door and quarter panels are painted sheetmetal, which was painted in two different ways: It was painted the color of the upholstery (i.e., red metal, red door panels), painted with interior paint that contrasted with the upholstery color (i.e., black metal, parchment [white] door panels).

In 1968, Chevrolet changed the door and panel design to have the door panel wrap all the way up from the bottom to the window. This change probably helped speed up production because a smaller area of the interior needed to be painted. A lot of drivers and passengers like to rest their arm on the top of the door, so 1968 and 1969 owners enjoy the comfort of upholstery-covered door panel caps in hot climates.

Standard Interior Panels: Most aftermarket door panels do not fit like the original panels, so if you have original door panels that are in decent shape, either use them or find somebody that may be interested in having an original set (or at least the original parts off the panels).

On original 1968 and 1969 interior panels, the curved upper section of the panel features a steel reinforcement molding under the vinyl. To keep the cost down, companies do not typically reproduce this part, which is why there are so many reproduction door panels currently on the market using plastic upper sections. GM also moved the door lock farther back toward the rear on the 1969 door panels.

You can purchase interior panels in two ways: "assembled" with plastic top sections or "partially assembled." If you choose partially assembled, the face of the door panel is assembled yet there's additional fabric at the top of the panels so you can install your original sheetmetal upper moldings. Hence, you can completely cover the interior panel, which is not difficult to accomplish for moderately skilled individuals. The partially assembled panels require that you also install the stainless steel trim and upper weatherstrip for the window, which can be used from your original panels or purchased as reproduction pieces.

Original door panels have two pieces of vinyl covering them. A larger vinyl panel covers the lower vertical portion. The top curved section lies over the lower flat panel, which is held in place by a long stainless trim piece.

The original panels used panel board to create the definition lines, and the wiperstrip that rides against the window was stapled to the door panel. Reproduction panels are not created the same way and have some distinctive differences, which aren't a problem for most enthusiasts. These newer panels use one piece of vinyl for the whole panel. Reproduction companies also use foam-padded vinyl over their panels to give them the definition lines, which makes them puffy and thick. These panels are thick enough to cause interference with other interior parts in the doorjamb area. The kick panels leave distinct impression marks on the door panels.

The window felt strip on the top is typically riveted to the panel and is visible from the outside of the car. If you want to hide the rivet heads you can touch them up with a dab of black paint or with a black permanent marker (if the manufacturer has left them silver).

Deluxe Interior Panels: For the 1967 model year, the deluxe panel was only a partial panel and, therefore, the tops of the doors and quarters were painted steel. The 1968 and 1969 deluxe panels were much like the standard panels, which were vinyl-covered sheetmetal on the top portion.

Assembled reproduction 1968 and 1969 deluxe interior panels are much like the standard interior panels in the respect that the top section is plastic, instead of sheetmetal. Because the door panels are a more involved construction, compared to the rear quarter panels, they are only available as assembled. The rear quarter interior panels are a fairly simple construction and are available to be built with the trim and sheetmetal from your original interior panels.

Door Mechanicals

Before installing watershields and door panels, take extra time to assess the condition of the handle operations and door latch mechanisms, as well as window track and crank mechanisms. If you don't test all these parts before installing the watershields, there's a good chance you'll be taking them off again, which is sometimes difficult without ripping them, which renders them useless.

Interior Door Handles

The interior door handle for standard and deluxe panels should spring back into place once you've pulled it and released it. You need to replace the door handle mechanism if it doesn't snap back into place. (The spring in the assembly is buried under welded tabs that would require a lot of surgery.) The handle mechanisms (standard panel interiors) normally wear a bit making about 1/2 inch of handle movement to and from the door, measured at the tip of the handle. Any more wiggle than that and you should consider replacing it so you can rely on the door operating normally if you had to open it in an emergency situation.

Door Latch Mechanism

Often with first-generation Camaros, you can have problems unlocking the door with the key or the locking knob on the top of the door panel. After quickly turning the key back and forth or vigorously pulling the lock knob up, you'll eventually get the door to unlock, but this is a hassle when you're trying to get into your car in a hurry. By always locking the door with the key, you get around this problem until you have time to fix the spring in the door latch. If you lock the door with the locking knob, the latch resets itself and requires the annoying action to unlock it. You can use the key or door lock knob to unlock the door after using the key to lock it.

The door latch mechanism has a little spring that typically breaks after many years or after somebody uses a slim-jim to gain access to your car. You can purchase the spring separately from reproduction parts companies or you can purchase the entire mechanism. The spring by itself costs about 5 percent of the price of the entire mechanism.

You can service the original mechanism with some assembly grease on all the pivot points reachable with a cotton swab. In order to remove the three counter-sunk hinge screws you need a #3 Philips screwdriver that's in good shape. If you attempt to remove them with a #2 or a #3 with a damaged tip, you're bound to round out

If one of the internal springs breaks in the door-latch mechanism, you'll have to buy a whole different unit, since the assembly is completely welded together. This OER-reproduction unit came from Classic Industries.

the slots in the screw head. Once you round them out, you are forced to drill them out and replace them. Use the right tool for the job!

Door Windows

If the window crank post is completely loose or has a grinding feel when you rotate the crank, you need to replace the window regulator. There's a different regulator assembly for standard and deluxe interiors because the window crank post is shorter for standard door panels. With the window rolled halfway up,

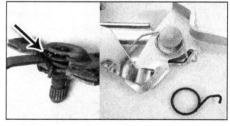

These two springs are necessary for proper door locking and unlocking operations. If you have to pull the knob or turn the key more than once to unlock the door, this inexpensive spring (right) is broken. If the interior handle (left) does not spring back, the mechanism should be replaced.

CHAPTER 12

you should be able to grab hold of the window with both hands and lift the window up to 1/4 inch. If you can lift it more than that, the rollers need to be rebuilt.

If you need to grab the quarter window and manually assist it in order to get the window to roll up or down, it's time to rebuild the guides and rollers.

Rebuilding a Window Mechanism

Over the course of 40 years, a Camaro's windows have been rolled

Window Mechanism Rebuild

1 If your window was adjusted correctly to start, use a fine-tipped permanent marker to indicate all the adjustable points so you can refer to them later. Once all the slack is removed, they won't be exact, but they may get you really close.

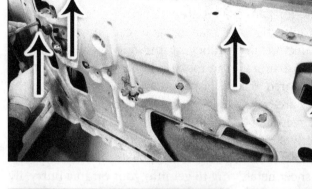

2 With the window up, remove the front and rear up-stops through the front and rear access holes. Loosen the two felt-faced sliders (arrows).

3 Put the window into a lower position, measure the depth of the cam bolt sticking out of the channel, and make notes. Unbolt the front and rear cam bolts from the channel.

4 Lift the window level, but tilt it inward slightly to get the rollers out of their guides. Then move the window forward slightly to miss the upper rear window felt-faced slider.

5 In order to remove the special glass-attaching hardware, cut up an old socket to make your own special tool or purchase the correct socket, like this one from Classic Industries.

6 Remove and replace all worn-out hardware. Check the guides and rollers, and make sure the correct plastic washers are between steel washers and the glass. Lube the guides, rollers and slides with white lithium grease.

7 Do everything in reverse and get the glass back into position. The two adjustable bolts on the track attached to the window move the window in and out. The slotted hardware on the top of the rear (not shown) and in the top front hole move the window forward and backward. You can change the upward stops at the front and rear. The OER window guides have new felt pads for smooth operation. If you need more instructions, check the Fisher Body Service Manual.

INTERIOR

up and down countless times, and as a consequence, the mechanical assembly inside the doors can wear out and fail. Replacing the slider, guides, and other hardware is not an extremely difficult procedure, but you must follow a few steps to complete the job properly.

Water Shields

A commonly overlooked detail is the watershield. This is the tar paper between the interior door panel and the door, as well as between the quarter panel interior sheetmetal and the interior panel. It is designed to keep water that gets into the door off of the door panel, which was originally made of cardboard. If you're thinking water would never fall off the glass and get on your door panel, think again. If this watershield leaks or is missing, the door panel gets soggy and moldy and you may get water into the passenger compartment under the carpet. If you've ever wondered how floorpans get rusty from the inside out, the biggest culprits were leaky or non-existent watershields or leaky window weatherstrips.

The interior side of the door and quarter panel has holes to access the door mechanism and window regulator hardware. These large holes allow water to get on the door panel if the water is not redirected to the long open drainage slot toward the bottom of the door and quarter. There are also little drainage indentations all the way down the sheetmetal.

The watershield has to be sealed to the door and quarter with a waterproof adhesive strip known as strip-caulk or dum-dum. It should be applied to the door and quarter around the water drainage indentations and should not have gaps. In addition, it should be placed around window and door handles and other holes to keep water from getting onto the interior panels.

All water should be redirected back into the inside of the door, where it drains out the small drainage holes at the bottom of the door. Tape the bottom corners of the shield to the door as the factory did. If your door has been hacked and has large holes for speakers, use some heavy-duty thick-mil plastic for a patch panel to direct water back into the interior of the door (as you did with the water shield). The factory also had a large round plug on the interior face of each door at the bottom near the door hinges. If yours are missing or broken, get new ones or add a patch over those holes too.

Reproduction watershields come in paper (as they did from the factory) with one black side, which is water resistant. Watershields are also available in clear plastic and are much more durable than the paper ones. The clear plastic ones allow you to see how well you sealed the

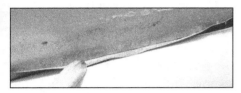

Here's the type of water damage that happens to a perfectly good door panel after a few rains if you don't have water shields installed. Water getting this far also gets into your carpet and floorpan where rust breeds.

The bottom is tucked into the long slot at the bottom of the door, and then taped, as they were from the factory. Add a patch to the panel if somebody cut a speaker hole in the door and/or if the plastic factory plug is missing at the front of the door.

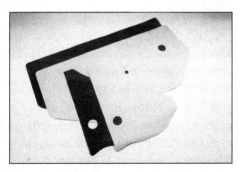

If you're installing reproduction paper watershields, they come rolled in a tube like a Camaro poster, so you should unroll them and lay them flat for a few days to help make installing them easier.

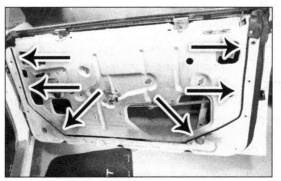

We installed clear plastic watershields because they are far more durable and it's easier to determine if they're providing a seal over the paper types. The strip caulk is around the entire perimeter of the water channel. Simply press the shield into the caulk to seal it.

HOW TO RESTORE YOUR CAMARO 1967–1969

CHAPTER 12

watershield to the door, and they allow easier future repair of door internals with less potential to damage the door upon removal. They also allow you to see what's going on inside the door if inspection is necessary.

Water Drains

We're covering water drainage systems in the interior chapter because two of the major drainage systems require the removal of interior panels.

A lot of design work went into getting water out of the car when it gets into the body, as well as keeping water out of the interior. There are drains in the rocker panels, at the bottom of the tulip panels (sides of the cowl panel), bottom of quarter panels (below the quarter windows), and at the bottom of the doors. In order for these drains to operate as designed, there must not be dirt and other debris clogging them.

Unfortunately, the factory didn't feel it was necessary to put more than the cowl panel with long wide slots over the cowl openings at the base of the Camaro windshield. These long slots allow all kinds of leaves and pine needles to fall into the cowl. These unfortunate gifts from nature work their way over the sides of the cowl inside the tulip panels and get trapped behind the fresh-air vents in the kick panels. If you combine 40 years of dust and dirt with a bunch of degraded leaves and pine needles, and then mix in a little water from rain or washing your car, all of the sudden the drain at the bottom of the tulip panel is clogged. The unwanted debris becomes damp and stays damp, which is a perfect environment for rust.

Whether or not you're doing a restoration, you should pull the kick panels out and clean out the fresh-air ducts and water drains. The kick panels are sealed to the metal around the fresh-air grille with dum-dum to keep water from leaking into the interior behind the kick panel. Make sure you remove all the screws from the panel before trying to detach it. If you have a 1969 with the separate grille, make sure you remove the screws behind it. On 1969 models, make sure you also remove the vent cable from the upper vent. These cables are very fragile, so be careful not to break them.

The easiest technique for removing the kick panel is to pull it away from the side of the car with your fingers behind the panel and against the firewall around the fresh-air duct. Once the duct is loose from the hole in the side cowl, the rest of the removal requires maneuvering the doorjamb area of the panel, and pulling it the rest of the way from the side cowl. The only way to remove the driver's-side kick panel is to remove the parking brake assembly first.

With the kick panels removed, clean out all the debris. Depending on the status of your car (driver or been parked for years), it's a good idea to wear gloves to protect your hands from a spider, mouse, rat, or sharp rusty edges. Once the debris is removed, use a piece of wire hanger to clear out the drain at the base of the cowl. Dribble some water down in there to get out the rest of the dirt and degraded leaves. If you really want to

This car will be driven a lot, so we added a Thermo-Tec heat-and-noise-suppressor mat to make the ride more enjoyable. We laid it out right over the KBS after it had fully cured. We were careful not to get the sticky product in fastener holes where it can gum up the hardware. To ensure it is securely laid down on the floorpan, a small handheld wooden roller was used to press it down into the contours of the floor.

The strip caulk is applied to the vent area on the kick panel before installation, so that it can keep the water from getting into the car through the fresh-air duct. The caulk stays pliable. Don't use hardening sealant or the kick panel will not come off again without destroying it.

We went back over the area with another coat of KBS RustSeal and BackBone Reinforcing Mesh for a much stronger rusted area, and then we used the KBS RustSeal on this floor pan that had an extremely small section of rusty pin holes (not enough to warrant panel replacement). A little KBS NuMetal epoxy is placed under the car.

INTERIOR

Originally, the carpet was designed to be installed over the factory tar-and-cotton underlayment. The jute padding is attached to the carpet to help fitment. If you cut the carpet too short, add more jute to make fitment easier. To flatten out the underlayment, heat it with a hair dryer. The Thermo-Tec is thin enough that it didn't affect the carpet.

If your he carpet arrives rolled up in a box, find room to lay the carpet out for a few hours to inspect it for unnatural warpage. Do not trim or cut the carpet before you put it in the car. This is a good time to take care of last-minute details that may have been overlooked (console nutserts, body plugs, shifter holes, etc.) before installing the carpet.

Use an ice pick or an awl to find the holes for the seats, seat belts, and other accessories. For anything larger than a trim screw, cut a hole in the carpet for the bolts. If you don't, the bolt can snag the carpet and unravel it, making a huge mess as you tighten or loosen the bolt. After you've poked the hole in the correct location, roll the carpet back and cut a hole in the backing just about 1/8-inch larger diameter than the bolt hole. Cutting an X instead of a hole is a sure-fire way to snag the carpet during installation of the bolts. Don't forget to cut the hole for the flasher grommet.

protect your investment, also clean out the inside cowl panel plenum.

Take a small brush and clean all the seams you can reach. Rinse everything and dry the area with a towel and/or compressed air. Use KBS products (or some other rust-preventative paint) to coat every inch inside the side-cowls and the inside of the cowl plenum and seal all of the seams with seam sealer, except the drains. To further protect the seam sealer, add a coat of gloss (KBS topcoat or another brand) paint to give water a smooth surface, so it exits the drain faster in the future. The factory never spent this much time to seal these water drains and fresh-air passageways and that's how these inner panels and cowl plenums have become common rust areas over the years.

When reinstalling the kick panels, don't forget to clean off the old sealer and put more dum-dum on the panel around the base of the vent protrusion. The dum-dum is a non-shrinking waterproof sealant that stays pliable (like clay). You need to use it when reinstalling the kick panels to get a secure seal. Never use any kind of silicone-based sealant or cement in these areas. (That would probably glue the panel to the steel and you would need to destroy the panel in order to remove it in the future.)

To help line up the screw holes when reinstalling the panel, stick an awl or an ice pick through the existing hole in the kick panel into the hole in the metal side-cowl.

The doors and quarter panels have drains for all the water that gets into the panel through the window and the side-glass weatherstripping. When you've got the door and quarter interior panels off, you should clean out the drains at the bottoms. You'd be surprised at how much junk gets inside the body panels. There could be a lot of small bits of glass in the doors or quarters if your side glass has ever been broken, so be careful. Clean and rinse the drains the same way as suggested for the cowl areas. Dry them completely and treat them with KBS products (or some other rust-inhibitor product), seam seal, and paint as instructed for the cowl areas.

Carpet

Original floor covering was 80-percent rayon and 20-percent nylon looped carpet molded to fit the contours of the floorpan. Made of two pieces, the rear half is installed first and the front section is installed over the top of the rear half. The factory carpet seam is directly in front of the front seats. Reproduction carpet manufacturers don't always get the seam location correct, which isn't a big deal for most people.

With the front seats installed, you should be able to lift the front carpet without unbolting the seats. In other words, the carpet lays directly over the front-seat bracket feet, not under them. You can get carpet in the original 80/20 loop style (carpet loops stick up) for a classic look or cut-pile 100-percent nylon (frayed strands stick up) for a custom look. Ken Howell of Auto Custom Carpets says that the loop carpet is correct for the restoration, but the cut-pile lasts longer and is what all the new cars use.

Most carpet kits come with the additional small sections of jute padding attached to the back side of the carpet as they were from the factory.

Dash Pad

Unless your Camaro has been well maintained for its whole life, the dash pad probably needs to be replaced. Over the years, it gets sun-beaten and develops cracks from merely touching it. The factory dash pad is made of a sheetmetal foundation that's covered with vinyl over foam. Certain aftermarket dash pads have been of notoriously bad quality (compared to original) and fitment. They are produced of a harder urethane than the stock dash, even though the stock one has hardened over the years.

OER started producing a much nicer dash pad than its older design. The newer-design OER dash could be developed because the more reasonable cost of newer technology has made it possible to create a completely new dash pad with better materials. Make sure you're aware of the differences before ordering yours. Al Knock Interiors also produces dash pads for Camaros, so check its products too.

If you want your original, aged dash pad rebuilt, Just Dashes is one of the sources that can reproduce 1967 and 1968 dash pads, as long as you send your original pad. The process is more expensive than good reproductions, but customers have been happy with the results. Even though its name implies otherwise, Just Dashes rebuilds and repair just about any padded or hard-plastic part of the interior.

To remove your 1967 dash pad, you have to remove the sheetmetal nuts under the top dash panel and the screws from the front.

The procedure with the 1968 dash is the same, but it has additional hardware and a clip on the end of the legs that drop off on both ends of the pad. Each dash pad leg has a post that sticks out and is pushed into a clip in the sheetmetal dash. Use pliers to pinch the clip on the top and bottom, and the clip pops off the pad and out of the dash. Or, you can stick a screwdriver in the clip from the back of the dash and pry the clip open.

The 1969 dash has clips that slide into the top of the dash and are held in with sheetmetal nuts and screws. They are a little easier to remove if you can gain access from behind the dash.

Console

The first-generation Camaro center console is cast-in plastic, with a few metal brackets, trim, and doors. It's not meant to be a seat or a foot stool. A common part to break on all first-generation consoles is the plastic ashtray door pivot points.

The weakest part of the 1968 and 1969 console that seems to be broken a lot is the plastic between the console door and the ashtray door. Unfortunately there's no easy way to repair the broken plastic and have it hold up to regular use once it's been broken, especially if you're needing to make a repair that disturbs the Madrid grain. Luckily, OER reproduces them, and most the parts sources sell the console in parts or as complete kits.

The 1967 console was a one-year-only piece for the Camaro, and the 1967–1969 Firebird (with different trim on the top) shared the basic console. The 1968 and 1969 Camaro console was not used in any other production car. If your car didn't originally come with a center console (RPO D55) and you want to install one, you should know that it has an interior light and other wiring (depending on transmission type and optional gauges).

If you're installing a center console in a 1968 or 1969 Camaro that never had one before, find the small dimples on the top of the transmission tunnel that show where to drill the holes to mount the console. There are a few extra dimples, so set your console on the tunnel to identify the right holes. There are two of them, side by side, in front, which are the exact width of the console bracket.

The rear mounting hole is the dimple in the tunnel that's about 32 inches back from the two front holes. Once you've got the rear dimple drilled you can lay the console

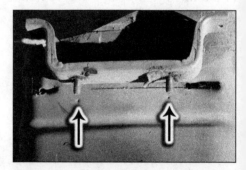

The transmission tunnel has dimples (shown here on a tunnel) to locate your drill bit in order to drill the holes for mounting the center console. Be careful though, there are a couple more dimples than needed.

Here's what the seat looks with the cover off the foam. The original seat foam breaks down over the years; the driver's side gets the most abuse.

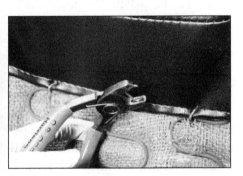

To lock the cover to the seat frame, use some hog-ring pliers and hog rings. These pliers are available from Eastwood in straight or angled versions. Lock the fabric in the same places it was connected originally.

Now the seat looks as good as new. There shouldn't be any bulges or bumps in the fabric at this point. If there are, you need to address them because they won't typically go away on their own.

This was an exclusive option for 1969—"Comfortweave" upholstered seats were available with Deluxe interior panels. The bordering material was the standard Madrid-grain vinyl with the special inserts.

Starting in 1968 you could purchase the ever-popular houndstooth seat inserts when you ordered the Deluxe interior. In 1968 you could only get Houndstooth with black or pearl parchment vinyl borders, but in 1969 you could order additional borders and Houndstooth colors.

This rare triple-green Berger RS 427 COPO Camaro originally belonged to Dale Berger's wife. The interior has standard buckets, standard interior panels, 140-mph speedometer, console, and Z23 wood-grain accents on gauge cluster, wheel, and center console.

on the tunnel and confirm the center mounting bolt hole is drilled in the proper location. The two front holes require nuts to be attached to the bracket from under the car, and the other two require nutserts to be mounted to the tunnel.

If you're installing a console in a 1967 that didn't originally have one, you have to drill your own holes. The factory didn't lay them out with dimples in the floor for you. The console mounts in a similar fashion to the 1968 and 1969 console, but not exactly, so pick up a copy of the 1967 *Camaro Assembly Manual* for all the proper measurements and schematics.

Seats

The factory seats were available in many different colors and materials. There was a standard interior for coupes and convertibles with a vinyl bench front seat (except 1969) or vinyl bucket seats.

The deluxe interior seat package featured a front bench seat (except 1969) or front bucket seats, all with at least the Madrid-grain border material, but there were changes for each model year. In 1967, the deluxe seats had all the interior colors available with a contrasting accent band. In 1968, the deluxe seats came with inserts that were color matched to the borders. Also available in 1968 was the black-and-white houndstooth pattern available with black or pearl parchment borders.

CHAPTER 12

For 1969 only, Chevrolet offered a new fabric named "Comfortweave" that was included with deluxe interior panels. It was color matched with the border color, available in black, red, blue, Medium Green, and Dark Green. Chevrolet did away with Madrid vinyl inserts, but offered four Houndstooth color options: black and white with black borders, black and white with ivory borders, black and orange with orange borders, and black and yellow with yellow borders.

Headrests

Today, the law requires installation of headrests in an automobile because they keep your head from whipping backward in a rearend collision, which can at the very least give you whiplash. Well, in 1967 the headrest was a $52.70 option (RPO AS2) and only 2,342 (of 220,906 1967 Camaros sold) left the factory with headrests. In 1968, AS2 headrests were on 2,234 (of 235,147 Camaros). New regulations for 1969, which took effect January 1, 1969, drove Chevrolet to add headrests in all 1969 models, but the 1969 Camaros could be special ordered with deleted headrests for an additional cost.

With production numbers so low in 1967 and 1968, it's easy to understand why you don't see many headrests. If you do have a set of seats from those two years, you should hold onto them or sell them to the highest bidder, no matter what shape they're in.

The 1967 and early 1968 headrests were the tallest of the first-generation Camaros and can be identified by their 7-inch chrome bar (it's the shortest of bars of all in headrests). Then Chevrolet switched to a wider and shorter headrest for the rest of the 1968 model year.

In 1969 Chevrolet slightly redesigned the headrest, but not by much, and it can be identified by the post length (10 inches in 1968 and 11 inches in 1969). Two different headrests were offered. The earlier version was in exact alignment with the seat, which had a slight backward slant. The later 1969 versions were angled slightly forward. The headrest post is the part that was bent and differentiates the two. The posts have their build dates etched on them for 1969, but earlier headrest stamps were not actual dates.

Steering Wheel

The original factory plastic steering wheels are typically cracked and need a little repair or replacement. A couple of companies, such as Eastwood and KBS Coatings, sell products to repair the plastic steering wheels. Although I haven't repaired a cracked steering wheel, it is a viable option. Apparently the keys to good results are initial preparation, a good strong product that will not shrink when it dries, your skill level to make the repair undetectable, and the proper paint that will stand up to the abuse of being handled while driving.

If you don't take on the repair yourself, you can purchase a steering wheel (new or good used) and send it to a shop that specializes in steering-wheel repair. Keep in mind that any restoration done by adding small sections of compound to large cracks, and

This 1969 RS Camaro has a smooth band between the two grooves instead of the distressed leather grain. During the 1969 model year, there were two similar steering wheels available with the same option designation. (Photo courtesy Tony Lucas Collection)

This is a 1969 427 Yenko (note tach location) with standard interior. The center shroud and horn buttons have a pebble grain and the band between the two grooves on the face of the steering wheel has a distressed leather grain. (Photo courtesy Tony Lucas Collection)

INTERIOR

Earlier in 1969 there was a walnut wheel, which was replaced with this Rosewood steering wheel about halfway through the model year. This SS/RS 396 has the 120-mph speedometer, 7,000-rpm tachometer with 6,000-rpm redline, and clock. (Photo courtesy Tony Lucas Collection)

Seat Belts

As of this writing, all but the state of New Hampshire have some sort of law to enforce the use of one of the most important safety devices in a vehicle, the seat belt. Whether you like it or not, the seat belt could save your life. If you have a stock first-generation Camaro, your car came

MCC's belts have all powder-coated stainless-steel attaching hardware and TIG-welded construction. The belts only attach at existing factory mounting locations and use high-quality hardware. (Photo courtesy Morris Custom Classics)

covering it with paint will never be as good as the original solid casting.

The best restoration is cost prohibitive for Camaro owners because it entails completely taking the wheel down to the metal and re-casting the whole wheel in one solid color. Quality Restorations does this for customers who have extremely rare steering wheels. A decent reproduction wheel is a fraction of the price of Quality Restorations' full-blown restoration or custom-diameter stock-appearing wheels.

By browsing Internet auction websites, you're bound to find a decent original steering wheel for sale. You can also look at all the restoration parts sources and probably find a reproduction steering wheel in the color and style that you need for your project. And if you're not interested in stock steering wheels, there's probably a tasteful and stock-looking steering wheel or two offered by an aftermarket company, like Grant Products' 14½-inch-diameter Classic GM wheels.

There were several different stock factory wheels for each of the three first-generation years. There was a standard interior version for each year with a Camaro emblem or Chevrolet emblem for the base models and RS and SS emblems for the same steering wheel. Completely different optional upgraded wheels were also available for each year with the same emblem treatments, from base to RS and SS. There was also one optional simulated wood wheel for 1967 and 1968, but in 1969 Chevrolet offered a walnut (early) and rosewood (late) steering wheel. In 1969 it also offered an additional rare-option soft-grip wheel.

You can get stock-appearing retractable three-point seat belts from Morris Custom Classics, LLC, for increased safety. Since these are available with all the factory colors and seat-belt-latch options, the seat belts don't look out of place to the untrained eye. (Photo courtesy Morris Custom Classics)

HOW TO RESTORE YOUR CAMARO 1967–1969

CHAPTER 12

Stock seat belts came in two types—standard and deluxe. The deluxe belt latches changed during the 1969 model year. The belt webbing was color matched to the interior as well as the latch assemblies on standard belts. Two different types of webbing were used over the three years.

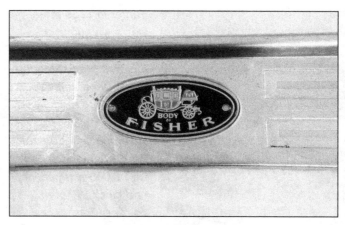

Here's an original unrestored 1969 sill plate with the separate Fisher Body carriage emblem riveted to it with extremely small-headed rivets. Reproduction sill plates don't have separate emblems or the rivets.

with a lap belt and a shoulder harness that tucked in clips in the headliner. We can't speak for everyone, but our experience has shown that Camaro owners these days rarely use the shoulder belts because they are tough to roll up and store in the factory-supplied hooks, and the shoulder belt is not self adjustable like today's belts. In fact, when replacing the headliner, most enthusiasts don't even reinstall the shoulder belts.

You can install racing seat belts with shoulder harnesses, but those are not street legal in some states because they don't allow you to move in order to check mirrors and such. Like the factory shoulder belts, they are either tight on your shoulder and uncomfortable, or they are loose and not doing their job.

A few companies have attempted to put newer seat belt functionality in older Camaros and other muscle cars. The best one on the market for quality and design is from Morris Classic Concepts, LLC. It has combined the lap and shoulder belt into a single three-point seat belt along with a retractable mechanism. The shoulder belt bolts to the factory shoulder belt attaching anchor plate in the headliner. Other manufacturers of these type belts rely on a single bolt for the upper attaching point, but MCC uses both original anchor points.

MCC's kits come with all the necessary hardware and come in a wide variety of colors to match most interiors. MCC hardware has all TIG-welded construction and all attaching brackets are powder-coated stainless steel. The latches on MCC's three-point belts are late 1969 Camaro-style stainless-steel latches with the "GM Mark of Excellence" button or a non-specific starburst button. MCC did its best to offer a superior product and one that appears as if it could have been designed by Chevrolet.

If you want to have your original seat belts restored and/or repaired, Morris Classic Concepts, Snake-Oyl, and several other companies offer correct color and design belt webbing to update your old stained, frayed, and unsafe original belts. These companies take your original belts and cut them apart, rebuild and repair your latches, remove the belt labels and remove any stains, rechrome your hardware, and reassemble all the parts with new webbing.

Sill Plates

Every time you get in and out of the car, your foot probably brushes across the sill plate. Over the years the plates take a lot of abuse and are rarely reinstalled because reproduction companies make very shiny sill plates that liven up any interior. Like a lot of reproduction parts, they are a little different than the originals.

The original sill plates had a separate oval Fisher Body carriage emblem riveted in the center of the sill plate. For the basic restorer, reproduction sill plates work fine without the separate emblem plate and small rivets.

For the restorer who wants to go the extra mile, you can cut the emblem out of an extra set of reproduction sill plates or take them from your original sill plates and epoxy them to another set of new sill plates. That sounds simpler than it is if you're going into the details of adding the correct rivets. The rivets are really small and can be lost in an instant even when you're careful, so be very careful. You can purchase new small rivets or carefully remove

158   HOW TO RESTORE YOUR CAMARO 1967-1969

INTERIOR

Plastic Part Repair

1 Scuff the back side of the corner and the panel around the broken corner with some 60-grit sandpaper. This gives the glue a good, coarse surface to bond to for strength. Don't sand the gap between the two pieces.

To perform this task you need 60-grit sandpaper to sand the back side, fiberglass mat, baking soda, some sort of Super Glue, and some Zip Kicker to speed up the process for extra strength.

2 With a steady hand, glue the corner back on with some Super Glue to hold the corner in place. Use the glue sparingly and try to keep it off the face of the panel to minimize the amount of time sanding the visible side of the panel.

3 On the back side of the panel, add a few small strips of fiberglass mat over the sanded plastic and the glue joint.

4 Sprinkle a little baking soda over the fiberglass (both are to add strength to the panel), but don't cover the screw hole or add so much material that you have to shave a bunch off to get the panel installed.

5 Dribble Super Glue over the fiberglass and then spray Zip Kicker on it. The Zip Kicker causes a vicious chemical reaction with the glue, and there will be a lot of crackling and vapors.

6 Sand down the back side of the panel, if necessary, so you can install it. None of your repair area should be thicker than the highest ridges of the original panel; if it is your corner will probably break off during installation. If you were successful in keeping the glue off the front of the panel, you may not need to perform any bodywork, and the crack may be almost invisible. If the repair is visible, fill, sand, and paint to restore the finish to your liking. We've done this repair on a daily driver and after 13 years of service, the repair has not broken again.

The corner of this dash was repaired 13 years ago and is still hanging strong. Just be sure to make the repair no thicker than the dash bezel so you can tighten it down properly. Add a felt washer to reduce dash squeaks.

After years of working on modern cars, we have a tip for reducing interior squeaks, especially from the dash panel: Cut a small square of thin felt and place it between the metal dash and the plastic panel. the old ones and epoxy them onto the new sill plate.

Repairing Plastic Parts

For a broken interior piece like a dash gauge panel or console, there is a bonding process that should repair it and add some strength back to the item. It's not a great solution for all plastics or surfaces, but it's fairly strong if done properly. If you use this method to repair a console, don't expect the repair to be bullet-proof and use your console as an extra seat. Always use this method on the back side of your parts because it can be quite messy.

The best example is the gauge bezel on the 1967 and 1968 Camaro. It's common for people to overtighten the upper mounting screws or remove the lower screws (or they fall out). The upper screws hold all of the gauges' weight, which leads to at least one upper corner cracking off. If you're lucky and still have the corner, you can save the bezel by reattaching it.

To use this method, you need coarse sandpaper, Cyanoacrylate (Super Glue, or some other brand of this awesome boding agent), fiberglass mat, baking soda, and Zip Kicker. You can get Zip Kicker in a spray or liquid at most hobby stores that carry radio-controlled cars; it's an accelerant for Super Glue that's safe on plastics. If you thought Super Glue dried dangerously fast, wait until you spray some Zip Kicker on it. Before performing this process on your panel, we recommend doing a test project, so you know what to expect before tying it on your important parts. Perform these steps in an open-air environment due to dangerous chemical vapors and wear safety glasses.

CHAPTER 13

MISCELLANEOUS MECHANICALS AND OTHER ITEMS

This chapter covers mechanical parts and items that didn't seem to fit into previous chapters. We've attempted to include most of the systems and aspects throughout the book, but some information might be too detailed or too trivial to include because of space constraints. Some information was simply left out because it is readily available on the Internet

Because this book is meant to be a practical guide to restoring Camaros, we've included some very detailed information about correct factory-original restoration tips and finishes. We've also included information on non-original restorations for the enthusiasts looking to build a nice driver but don't care about such things as correct date codes on water pumps and the correct finish on a hood hinge. Some enthusiasts are more interested in improving their cooling and fuel systems.

Cooling System

In general, the Camaro's cooling system was similar to most Chevrolet cooling systems in late 1960s passenger cars. When it comes to the details

The big-block Camaros, like this 1967 427 Yenko, have a distinctive heater core that sticks out through the blower motor duct on the firewall, instead of being wedged behind the passenger-side valve cover like the small-block and L6 heater cores would be. (Photo courtesy Brian Henderson)

of different options and engine sizes, there are way too many differences to cover all three years, so we highlight some details regarding the cooling system. We also cover some basics and some information regarding some modifications you can make to improve upon the stock cooling system.

Fan Shroud

The fan shrouds for big-block and small-block engines were different and the same goes for engines with and without air conditioning.

Find the right fan shroud for your car and install it. That seems really basic, and it is.

Without a fan shroud, you risk injury and perhaps your life because a loose piece of clothing can easily pull you into the fan. The shroud does more than protect your well-being; it also helps the fan pull air through the radiator. Without a shroud, the fan pulls air from all around it, but fails to pull air through the radiator. When driving down the road at more than 20 mph, air forces its way through the

MISCELLANEOUS MECHANICALS AND OTHER ITEMS

Restoring a car back to factory original involves many details. However, new original information comes out periodically. Here, we see a 1969 L72 with the correct square and rectangle cuts in the top of the radiator (blue arrow), but the radiator is not glossy enough. The fan shroud is a 1969 NOS part, but the NOS is different than factory original because this one has the long bumps/humps on top of the shroud (black arrows).

radiator, but when you're sitting in traffic or at a stoplight there's nothing to make sure air is pulled through the radiator if the shroud is missing.

For the best performance of the fan and shroud, the fan diameter should be no smaller than 1½ inches smaller than the diameter of the shroud. If you're going to install an aftermarket fan shroud, it should resemble the stock fan shroud. If a shroud doesn't fit against the radiator, it won't pull air through the core and you're wasting your money.

Water Pump

How do you know if your water pump needs to be replaced? The front seal is the most common part to fail and when it does, coolant leaks out of the small hole in the front of the water pump below the shaft. This hole is referred to as a weep hole.

The shaft bearing can also wear out, and when it does, the front seal follows suit. If you're going to restore your Camaro back to completely stock, you need to have an original water pump. Over the years, water pumps have become a throw-away part instead of rebuilding them due to the necessity of using a hydraulic press and shop tools to remove and re-install on the front hub in order to replace the shaft seal. Original cast-iron pumps with old worn-out bearings and worn-out seals were either tossed in a metal scrap pile, traded in as a core, or thrown out with the garbage. These old pumps can be rebuilt with the right parts and equipment. Rebuilding a non-original pump is going to cost you more shop time than it's worth. You should simply buy another pump.

However, if yours is an original with the correct casting numbers and build date, you may want to save it and have it rebuilt. Brian takes his original water pumps down to his local NAPA auto parts store and has them rebuild the pump for him on-site. This is not a service available at all NAPA stores because they are franchised and each store is different. Check with your local NAPA, auto shops, or services available by shipping your pump.

If you're modifying your engine to boost the performance, take into consideration that producing more power produces more heat, particularly for big-block engines. The only way to reduce heat is to get the hot water flowing out of the engine into the radiator, where the air cools it by dissipating heat before the cooled water completes the cycle and travels back into the engine to pick up more heat. If you're more interested in performance rather than having the factory-correct water pump with the cast-in numbers, Milodon and Stewart offer a few high-volume pumps available in cast iron or aluminum that don't look too custom and could be painted for a factory appearance. Some builders go as far as to get an aftermarket pump, grind off the company logos, re-finish the surface, and paint them to look factory.

With the exception of ZL1s, the original factory water pumps were painted after they were installed on the engine at the engine assembly plant. The pump was painted Chevrolet Orange with the rest of the engine. The ZL1 pumps were painted gloss black before the engine was assembled.

There are two versions of small-block and big-block water pumps. In 1967 and 1968, the small-block and big-block used short water pumps and had the alternator hanging way out to the driver's side over cylinder number-1. In 1969, Chevrolet switched to the long water pump and was able to move the alternator much closer to the centerline of the water pump on the passenger side in front of the cylinder head. This is a much better design because the shorter belt is much less likely to jump off the pulleys at higher RPM.

There are two different water pump shaft sizes. Standard-duty water pumps have 5/8-inch-diameter shafts; the heavy-duty pumps, usually reserved for Corvette engines, have 3/4-inch shafts. Some aftermarket heavy-duty pumps have 3/4-inch shafts, but the end of the shaft is reduced to 5/8 inches. Pay attention to the shaft size when ordering a pump because the hole in the center of the fan and pulley are hub-centric and need to match the size of the pump shaft. If you're upgrading from a standard-duty pump to a heavy-duty pump with the larger

CHAPTER 13

shaft past the mounting flange, the shaft needs to be modified to fit in the fan.

Radiator

The factory radiators were painted gloss black. Radiators are a "wearing" item, so after 40 years in service they've typically been replaced or rebuilt and repainted. After a repair, radiator shops usually repaint the radiators flat black, which explains why most original and replacement radiators are flat black.

Often, the radiator ID tag gets lost in the process of a shop repair. Factory radiators had an ID tag clipped to the tank solder joint. The tag has small part numbers, radiator configuration info (group of six letters and numbers), and the two large letters that designate the application. There are a couple of letters stamped in each factory tank that designate more information on the radiator and the vehicle application. The driver's-side tank has specifics about the construction of the radiator, and the passenger-side tank has letters that designate engine and transmission application. If you have a tag and/or the original radiator tanks, go to www.Camaros.org and compare your findings to the application data they've collected over the years.

As with any car, the Camaro was available with manual and automatic transmissions. Each of which required a differently configured radiator. Automatic transmissions operate off fluid that becomes extremely hot, so they need additional cooling. The solution was to have a transmission cooler integrated inside the passenger-side tank of the radiator. The manual-transmission-equipped cars didn't have a fluid cooler.

The stock radiator is made of a copper core and brass cooling tanks. This old construction is effective, but not as efficient as the newer-designed aluminum radiators. If you're considering the upgrade to aluminum, a couple of companies manufacture bolt-in replacement radiators. The only aluminum replacement radiator on the market that's not made with boxy fabricated tanks is the Griffin radiator. Griffin's tanks are stamped aluminum and, even though they don't look just like the stock copper/brass radiator, they do look the closest out of all the other aluminum radiators. There's an old rumor about Griffin radiator cores being epoxied to the side tanks. This is not true. That incorrect information was probably started by some competing radiator company. Griffin's cores are welded to the side tanks and then they add an epoxy over the weld to reduce problems associated with expansion and contraction of these welded joints.

Heater Core

A heater core is a small version of a radiator which has engine coolant running through it. In order to get heat into the passenger compartment, there must be a heater core located somewhere in the car where air runs across its external surface and the air is heated and is forced into the passenger compartment through vents located under and on top of the dash. In order to cycle water through the heater core, there is a pressure hose connected to the intake manifold where cycling coolant is trapped by the thermostat. The trapped water flow forces water through the heater core feed hose into the bottom connection of the heater core. The hose connected to the outlet (upper port) of the heater core is connected to the water fitting connected to the water pump, which helps pull water through the system.

The small-block V-8 (and L-6) non-A/C Camaros use the same heater core, which places the inlet and outlet hoses closer to the center

Original radiators had these tags clipped on the tank joint on the passenger side of the radiator. This one has a "UY" stamp, which designates it as a special HD radiator for 396 and 427 COPOs equipped with manual transmissions. Also notice the tank code "OO" stamped on the tank ribs. An original radiator with one of these tags is hard to find.

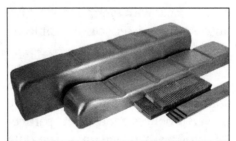

If you are going to upgrade to an aluminum radiator, the only company making units with stamped tanks is Griffin Radiators. All the other aluminum radiators on the market use fabricated aluminum tanks because they're less expensive to produce. At least the stamped Griffin tanks look closer to stock tanks. (Photo courtesy Griffin Radiator)

MISCELLANEOUS MECHANICALS AND OTHER ITEMS

of the firewall and to the right (if you're standing in front of the car) of the blower motor duct on the firewall.

Because the big-block engine is wider than the small-block V-8 and L6 engines, it's almost impossible to connect heater hoses to a smaller engine's heater core. Note that it was mentioned as being almost impossible. If you have short, original valve covers it's a pain, but they can be connected if you use molded heater hoses for late-model cars with 90-degree elbows on one end. Chevrolet designed a separate heater core for big-block Camaros. The heater core inlet and outlet are on the opposite side of the heater core, so that they protrude through the center of the blower motor duct, about 10 inches closer to the passenger-side front fender. This gives more than enough room for the big-block engine.

The heater core for an A/C-equipped Camaro is completely different than the ones for non-A/C cars and requires a lot more work due to the huge bulky A/C case that mounts to the firewall. The large A/C case requires removal of the fender in order to remove the exterior case. Therefore, in order to replace the heater core, you have to remove the case. That makes replacing the heater core on an A/C car a huge job.

Before installing a new heater core, make sure it holds 16 pounds of fluid pressure (the typical cooling system pressure). We've seen heater cores leak when you pull them right out of the box. If you don't have a way to safely apply 16 pounds of pressure, at least blow into one tube and hold your hand over the other to create a little pressure and check for leaks. When installing the hoses after the heater box and blower motor duct is installed, make sure you don't damage the heater core tubes and cause a leak.

Blower Motor Duct

If you are attempting to replace the heater core with the fender on the car, you have to remove the inner fenderwell first. The same goes for replacing the heater blower motor. In order to remove the small-block heater core, you have to remove the blower motor ducting on the engine side of the firewall. The small-block heater core tubes are curved and nearly impossible to remove on their own. If you do accomplish this feat without removing the outer blower motor duct, there's a big chance you had to tweak the brass tubing and possibly caused a leak, which would require doing the whole job again.

If you're converting your small-block or L6 duct to a big-block setup, you can purchase the correct heater core tube-seal retainer plate, screws, and gaskets and drill your own heater core holes to perform your own conversion or you can buy a whole new duct. Plus, you obviously need to purchase and install a big-block heater core.

To swap the heater cores, there's a different set of spring-steel clips that hold the heater core inside the heater box. The only place we could find that sells these special big-block heater clips is Heartbeat City. The small-block clips are just not big enough and are not shaped correctly to get the job done. You need to use one of the large seals included in your seal kit to fill the holes in the firewall where the small-block core tubes used to be. In order to seal the blower motor duct to the firewall, you need some 3M Strip Caulk to fill the gap between the two. Use Strip

Here's an original blower motor and duct from a 1969 Yenko. Part of the duct itself has been stuck in a time capsule under the fender. The 12 o'clock position doesn't look too glossy, but there's a mirror-finish reflection at the 3 o'clock position. The entire duct was gloss black from the factory. The part of the duct not covered by the fender was spray painted semi-gloss black from a poor restoration attempt. (Photo courtesy Tony Lucas Collection)

Caulk because it does not harden and crack, letting air and moisture into places you don't want it. Using urethane makes an ugly mess and you'll have trouble removing it if the heater core goes out again.

The factory finish on the duct is gloss black. It's one of the very few items under the hood that actually came from the factory with gloss black paint.

Heater Blower Motor

Replacing the blower motor is no small feat if the car has the fender and fenderwell installed. The inner fenderwell has to be removed. Before replacing a suspected blown heater motor, check the fuse and then the wire connection. Use a multimeter to confirm 12 volts are getting to the motor. Original blower motors have the date stamped on them as well as Delco Products. Also stamped is Rochester NY USA or Dayton Ohio

USA. Reproduction units don't usually look exactly the same or have USA stampings, which isn't a problem for most. The factory finish on the blower motor is semi-gloss black and the small mounting hardware is black oxide.

Cowl Hood

The cowl induction hood (RPO ZL2) was first released into production on 1969 Camaros. The cowl hood was called the "Super Scoop" in early 1969 Camaro advertising and installed on approximately 10,000 of the 243,085 Camaros built in 1969.

Most cars with a hood and cowl that meet a windshield with varying angles like the first-generation Camaro have a high-pressure area at the base of the windshield, which is why most cars from this period drew fresh air for the interior of the car from this area.

On a related note, that's also why GM pulled external air from this area to feed the cowl induction hood. An electric-solenoid-operated door was installed on the back of the hood opening to keep the door closed except under power when the throttle was opened to almost full-throttle. When the door opened, air was forced into the engine that was cooler than the stagnant hot air located under the hood and produced additional power.

Many people have installed these cowl hoods on 1967s and 1968s that were never equipped with the cowl hood, and on 1969s that did not originally have them. Unfortunately, these people don't also install the door and don't have the original air cleaner or correct underhood bracing to seal the air cleaner in order to fully utilize the pressurized air to boost

The Cowl Induction hood did not go into production until the 1969 model year. The bottom of the hood has additional bracing and a duct that seals it against an open-top air cleaner. The external air only enters the hood scoop when the flap on the back of the hood is opened by a solenoid operated by the carburetor linkage. Not having this flap closed is hazardous to your health.

performance. The problem with not having all the factory parts and baffles installed is that air entering through the grille passes through the engine compartment and goes out the back of the hood scoop where it goes directly into the fresh-air vents and into the passenger compartment. Not only does this make the air in the car hot, it also allows any burning oil, gas fumes, or carbon dioxide that may be leaking from an exhaust pipe to get directly into the car, which is very unsafe to breathe.

So if you are driving a Camaro with an improperly installed cowl hood and you constantly get headaches or feel bad after a long drive, you may want to consider sealing off the air cleaner or not opening your fresh-air vents. Even leaving the vents closed, however, won't completely keep the fumes out of the interior of the car because outside air easily makes its way through the heater system unless you've replaced all the 40-year-old seals.

Fresh-Air System

The only way to get fresh air in your face on the 1967 was by opening the vent "wing window." These little windows are a love/hate item—some love them and some hate them. Well, hate is a strong word; maybe dislike is better. The problem is they make it easier to break into the car, since the factory locks are not exactly foolproof, or some thieves simply break the little window to gain access to the car because it's less conspicuous. Because the 1967 had the vent wing windows, it didn't really need the vents in the face of the dash, so if you wanted outside air in your car it was either open one of the six moving windows or open the vents located in the kick panels. The 1967 did, however, have the dash vent balls and bezels in the face of the dash (same as the ones found on the face of the 1968 dash) on A/C-equipped (RPO C60) cars, and they were connected to the evaporator, not to the fresh-air vents as on the 1968s.

Classic Industries sells this whole 1967 and 1968 kit, which replaces the dash-vent bezels and balls as well as the seals and mounting tabs for the bezels. These bezels include the right and left bezels instead of the ill-fitting universal bezel that's supposed to fit both sides of the dash.

In 1968, Chevrolet discontinued the vent wing windows and added vents in the face of the dashboard as standard equipment (instead of only on A/C cars, as in 1967). This change was marketed as Astro Ventilation and continued with vents in the dash face but with a different dashboard in 1969. Outside air came through the face of the dash, the heater vents, and the kick panels—or by opening one of the four windows.

If you operated the heater in any 1967–1969 Camaro, the heated air blowing over the heater core (located in the heater box) was pushed out of the floor ducts on the top of the dash. It didn't come out of the vents in the side kick panels or the two vents on the left and right side of the dash face. The outside air entered the vehicle through the openings in the top of the cowl area, which is open to ducts that run down into the kick panels. At the bottom of the side panels, there is also a duct system that forces air down through the rocker panels to dry them out if water collects in that area.

Air Conditioning

We're only going to briefly touch on air conditioning because A/C is a very involved system of its own. There are two ways to get A/C in your Camaro. The factory system operates on R12, which is no longer in production, but you can still get your hands on it, or there are some R12 subsitutes that are supposed to work. There's good support for you if you choose to go with aftermarket A/C. Since this book is more about restoration than modification, this section is brief, but we're mentioning it because some enthusiasts prefer to install a system that has an additional 40 years of technology behind it and there's still support for the chemicals that keep it going.

Factory System

These systems are designed to use R12 refrigerant. The government passed legislation to keep the public from being able to purchase R12. The EPA banned production of R12 at the end of 1995. Apparently all the R12 that's currently available is what was stockpiled or recycled from air-conditioning systems that had R12 left in them. If you check the Internet you'll find places to purchase it for personal use. Note that the low supply and increasing demand has seriously driven up the price.

If your car was originally equipped with A/C but everything is missing, you'll have a tough time rounding up all the necessary parts. They're out there, because a lot of people have disassembled many air-conditioning systems and some Camaro owners have actually hung onto all the parts. There are new reproduction pieces to replace commonly broken or worn-out parts. OER has reproduced the evaporator case cover for the inside half of the unit for small-block and big-block cars. These are the covers that are closest to the engine and commonly become cracked from removing engines or broken in order to install aftermarket parts.

Aftermarket Systems

There are a few companies offering aftermarket air-conditioning systems, but the only one of those companies we've worked with is probably the oldest in the hot-rod air-conditioning business, Vintage Air. This company offers complete kits to install A/C in a Camaro that didn't have it, and it offers kits to

If you want to convert your original A/C system that used R12, contact Vintage Air. They have complete kits to convert factory A/C cars and non-A/C cars. The kits include everything needed to adapt your ducts—new glove box liner, dash control panel, compressor, etc. (Photo courtesy Vintage Air).

completely convert a Camaro that was originally equipped with A/C to a brand-new, more-efficient system.

The experience we've had with a Vintage Air kit to add A/C to a first-generation was good. All the necessary parts were included. Overall, everything went in well. We had to slightly modify a couple of the hardlines with a handheld tubing bender to get the condenser to fit the way we wanted, but considering all the work and parts that went into installing the system, it wasn't a hassle. If you want a truly custom and more hidden Vintage Air kit, give Detroit Speed (www. detroitspeed.com) a call and ask about what you need to install a Vintage Air Gen II Compac in first-generation Camaros. Detroit Speed also works closely with Vintage Air and makes some nice products that work together for an extremely clean install.

Carburetor

The first-generation Camaro came from the factory with either a Rochester carburetor for standard performance or a Holley carb if it was a high-performance-optioned car. CarTech has books on how to rebuild Rochester and Holley carburetors. If the thought of rebuilding a carburetor scares you, there are shops that specialize in rebuilding carburetors. Plenty of them rebuild very expensive and rare carbs for the pickiest restorers. Shops that come to mind are The Carburetor Shop (www.thecarbshop.com), Ace Fuel Systems (www.acefuelsystems.com), and Eric Jackson of Vintagemusclecarparts.com.

Fuel Line Sizes

If your Camaro was originally a 6-cylinder and you're upgrading to a V-8, be sure to check the fuel line size. The L6 cars came with a 5/16-inch fuel line; the V-8 cars (small- and big-blocks) came with 3/8-inch fuel lines. Most enthusiasts rebuilding first-generation Camaros are installing a V-8 engine with more than 350 hp. When bumping up the horsepower, the engine needs more fuel. The small 5/16-inch fuel line is not large enough to adequately feed the engine, and you could seriously damage your engine if it starts starving for fuel and running lean while driving under moderate performance driving.

A 350-hp engine needs at least a 3/8-inch fuel line. Some models with 225-hp small-blocks and up to the 375-hp (and 425-hp Dana 427) big-block-equipped Camaros had 3/8-inch fuel lines. It's fine that the factory used 3/8-inch fuel lines on higher-powered engines, but fuel requirements often exceed this fuel line size.

Engines with more than 400 hp in racing environments could require more fuel than a 3/8-inch line could effectively deliver, especially if you're using a traditional camshaft-operated fuel pump. These camshaft-operated fuel pumps suck fuel from the tank and push it to the engine. When the engine power level exceeds 400 hp, you should upgrade to a 1/2-inch front-to-rear fuel line; with more than 600 hp, you should use a 5/8-inch fuel line. These sizes change depending on the type and model of fuel pump. For more performance info, check some of the fuel system books on the market or consult the information that accompanies your performance fuel pump.

The factory offered mechanical block-mounted cam-operated fuel pumps for all applications. The original pumps were different for almost every application. Not only were they different in line size and fittings, they had different pressures and volumes too, depending on the application. Many replacement pumps are generic in their applications and configurations, which is okay for most enthusiasts. If you're installing a performance aftermarket pump, make sure you consider that a standard aftermarket street carburetor doesn't typically like more than 8-1/2 psi and stock carbs typically operate around 6 psi. Any more than what your carb is designed for and fuel starts pushing past the needle and seating in the carburetor, and the engine automatically floods.

You can purchase mechanical, racing fuel pumps with more than 14 psi, which require an external pressure regulator. For the factory-correct looking and fitting fuel pump for your application, you can purchase reproduction pumps from many restoration sources. If you're interested in aftermarket pumps that closely resemble an original, but aren't exactly correct, you can find a good one by Carter.

Installing a fuel pump on a small- or big-block Chevy engine can be a pain in the rear because the fuel pump pushrod always has a little bit of pressure on the lever arm on the fuel pump. It doesn't help that the rod wants to fall once you push it up into the block because of gravity working against you.

If you're installing a mechanical fuel pump on a small-block, there's an additional plate to install between the pump and the engine block. The big-block does not have this intermediate plate, except a handful of first-generation big-blocks produced in 1963 that are extremely rare. If you have one of these big-blocks, you've found yourself a goldmine.

Fuel Pump Installation

To install the fuel pump, follow these procedures:

Remove the threaded plug under the fuel-pump mounting pad from the block because you need to install the fuel-pump pushrod through that hole.

Put JGD assembly grease or moly lube on the tips and shaft of the pushrod to fully lubricate it. The extra viscosity of the JGD assembly grease is thick enough to help keep the rod elevated in the pushrod bore of the block.

Slide the pushrod up into the block through the hole under the pump mounting boss and install the threaded plug. If you're working on a small-block and the rod won't install in the hole, check to see if you've installed a bolt that's too long, in the boss in the front of the block. The bolt bosses on the front of the big-block are blind and do not affect the fuel-pump pushrod.

MISCELLANEOUS MECHANICALS AND OTHER ITEMS

Some engine builders tell you to install a bolt in the front hole to hold the fuel-pump pushrod up while you install the fuel pump, but if you accidentally put a nick in the pushrod or forget to replace the bolt with a shorter one, you'll damage the engine, so we prefer to use the following method:

With a 6-inch piece of coat hanger or stong bailing wire, hold the fuel-pump pushrod in its bore in order to get the fuel-pump lever arm under it so you can install the pump.

Before proceeding, spray the fuel pump gasket with a little Permatex High-Tack Spray-A-Gasket, or thin coat of Permatex Ultra Black, on both sides.

Stick two bolts in the holes of the flange on the pump, and then stick the gasket to the fuel pump flange. The gasket helps keep the bolts in place while installing the pump, but it won't do the whole job. You have to hold the fuel pump so that your thumb and forefinger hold the bolts in the flange.

Put a dab of JGD grease on the pump's lever arm.

Use the coat hanger to elevate the pushrod in the hole and slide the fuel-pump arm under the pushrod, and remove the coat hanger.

The fuel pump still does not easily fit against the block. You need to force the fuel pump upward a little to compress the arm of the fuel pump, so that you can install the fuel pump mounting bolts in the block. If the pump seems cocked to the front or rear, the pushrod will slip off the top of the lever arm. If this happens, use the hanger to get it back in the proper position.

Once you get the fuel-pump bolts started, there will still be pressure from the lever arm, making it tough to install. If the pump binds and it's not installed against the block, the pushrod has typically fallen off the top of the lever. (Or there's some other issue, such as a defective fuel-pump arm, the wrong fuel pump, or a fuel-pump rod that's wrong for the application.)

If you install the pump and pushrod, but the pump won't pump fuel, it's possible the pump actuator lobe on the cam is flat. The fix for this is to replace the cam or install an electric fuel pump. Small- and big-block fuel pumps may look similar because the gasket-sealing surface is the same shape and the fuel-pump arm looks similar, but the arms are not the same and the fuel pumps are not interchangeable.

In 1967–1969, small- and big-block engines were equipped with a solid-steel fuel-pump pushrod. The pushrod is operated by the rotating off-center round lobe near the front of the camshaft. The rod actuates the arm on the fuel pump, and in turn pumps fuel from the tank into the carburetor. The solid pushrod is a heavy reciprocating mass that can cause fuel-pump cavitation when the camshaft turns at high RPM because the mass of the rod can't rebound from the spring tension on the fuel-pump arm. ARP and other manufacturers make a lightweight aluminum-bodied fuel-pump pushrod with hardened caps on the ends. The lower weight reduces fuel-pump cavitation because the spring operating the pump arm has less weight to control.

The friction produced by the pushrod (stock or lightened aftermarket) being actuated by the cam and operating the fuel pump robs a bit of horsepower. This is why racers wanting an additional edge on the competition switch to electric fuel pumps. Less friction always means more power.

Hood Latches and Hinges

The hood hinges, latch assembly mounted in front of the radiator, and latch plate on the hood were all phosphate plated from the factory and were the same shade of gray. Even though it's really hard to tell on some of the most pristine original examples we've seen, we've always been told that the hinge parts were phosphate plated before they were assembled with the rivets. This would mean that the rivets aren't phosphate plated, which means if you are looking for the factory appearance you can remove the plating from the rivets—with a lot of tedious work.

If you want to do phosphate plating at home, consider that the modern phosphate color may be way off the factory color. Check around on the Internet to find a home kit

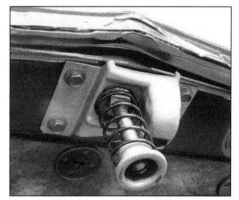

The hood latch pin was black oxide. Its cup and bracket were gray phosphate and the spring was painted gloss black. The hood-latch assembly that this locks into in front of the radiator is the same gray phosphate, and the bolts are black oxide. (Photo courtesy David Boland)

This 1967 Yenko hood pin has a threaded shaft, which threads into the original hood height adjuster threads in the core support. The hole in the hood goes right through the thick pad on the hood. Also note the correct radiator support semi-gloss black finish. (Photo courtesy David Boland)

Unlike the 1967 and 1968 RSs (Z22), the 1969 RS headlight covers were the first ones to have grilles that allowed light to shine through if the headlight doors failed to operate. Because this car has Tuxedo Black paint, it came with a silver RS grille. The small chrome parts above the headlight are headlight washers. The 1969 RS package came with a pair of headlight washer nozzles mounted above each headlight. Only 116 non-RS Camaros have the $15.80 option (CE1). (Photo courtesy Tony Lucas Collection)

The stainless trim ring around the headlight bezel is part of the (Z21) style trim group, which also included trim on the simulated louvers in front of the rear tires, bright taillight lens trim accents, wheel-lip moldings, tail-panel Bowtie emblem, roof drip-rail moldings, and black-painted body sills. The small triangle in the center of the headlight surrounds the "T-3," which is what the original headlights wore. (Photo courtesy Tony Lucas Collection)

that people have had success with to get factory-looking results or find a company that performs the plating process and gets the correct finish and color for Camaros.

The factory finishes on the hood latch pin is black oxide and the spring is gloss black.

Endura Bumper

The Endura front bumper (RPO VE3) was only offered on the 1969 and was painted to match the body color. It's a urethane-molded front bumper that closely fits the contour of the recess in the front fender extensions and lower valance. Its edges and peak are much more pronounced than the rounded chrome front bumper.

To install one on a car that didn't come with one from the factory, you only need to purchase the four correct Endura brackets and correct hardware in addition to the bumper. Don't forget to add flex-agent to the paint before matching it to your body color.

Caleb Arone's 1969 Camaro has the standard chrome-plated front bumper. The perimeter of the grille matches the body color. Some people incorrectly paint it black or silver. This car has been restored to stock, with the exception of period-correct 1970s racing parts, which include a rare Moroso hood, bolt-in Lakewood rollbar, bumper tow-bar, and a 427 with a Weiand 2-4 tunnel ram.

MISCELLANEOUS MECHANICALS AND OTHER ITEMS

The color in this photo is not wrong. You're looking at a dark green vinyl top. The 1967 and 1968 models came with only black or white vinyl, but the factory added blue, brown, and green to its offerings. (Photo courtesy Tony Lucas Collection)

Vinyl Top

Buyers didn't have much of color option when it came to the vinyl top on the 1967 and 1968 Camaro. Chevrolet offered either black or white. The factory changed the vinyl-top color availability for 1969. First it changed the white to Parchment and then added blue, brown, and green.

Convertible Top

I could write whole chapters on the convertible tops and features that separate them from the coupes, but we're limited on space so I'll just scratch the surface on chassis and driveability.

Frame

Because ragtops don't have the structural support of the roof panel, first-generation convertible floorpans have additional bracing under the car to reduce chassis flex. For ease of servicing the driveshaft, there's a removable brace that greatly increases the rigidity of the frame and is bolted between the convert-

Convertibles have additional bracing on the floorpan to strengthen the chassis in the absence of the roof structure. The center bolt-in brace helps tie both halves of the car together and is extremely necessary for rigidity. (Photo courtesy Tony Lucas Collection)

ible bracing. I've seen quite a few convertibles missing this important brace. There are additional braces in the trunk and on the floorpan under the front of the rear seats. The rocker panels were also thicker-gauge sheet-metal compared to the coupe rockers. For some reason some coupes have been found with thicker convertible rockers. It's assumed that this anomaly is a case of running out of coupe rockers on the assembly line and simply installing parts they had on hand.

With the lack of the roof structure, first-generation Camaro convertibles were never strong-performing cars, especially around corners. One could argue that the 1967 and 1969 were both Indianapolis Pace Cars and that the convert- ibles were available with big-block power. Our position is that if Chevrolet thought the convertible could really perform well on a road course or drive well around more than a couple of banked turns around Indy, it would have offered a Z28 convertible, which it didn't. The stock convertible is just too flimsy for more than spirited street driving.

To strengthen the convertible chassis, you need to install a thicker convertible brace and bolt-in subframe connectors from Hotchkis. These items don't require floorpan modification. If you're more worried about having a strong convertible chassis that feels as rigid as, or stronger than, a stock coupe, you can upgrade and weld in a set of Detroit Speed subframe connectors. The strength of Detroit Speed connectors comes from cutting the floorpan and welding the connectors into it as an integral frame structure, connecting the front subframe to the rear leaf spring torque boxes. Beware that some less invasive and cheaper weld-in subframe connectors hang down a few inches, reducing ground clearance and blocking future removal of the rear leaf springs.

The round cylinder is called a "cocktail shaker." Located at all four corners of the car, they add stability to convertibles. You can also see the additional convertible support brace above the cocktail shaker on this 1969.

HOW TO RESTORE YOUR CAMARO 1967–1969

CHAPTER 13

Cocktail Shakers

The convertible bracing throughout the car wasn't enough to make the car feel stable during normal driving. In an effort to reduce the shimmy and shake of the Camaro after removing the roof, GM engineers came up with cylinders filled with weights atop springs in viscous fluid. The aptly named "cocktail shakers" reduced the torsional twist of the chassis when driving over uneven roads. They weigh approximately 25 pounds each—with one at each corner of the car, that's an additional 100 pounds.

In the past, Camaro owners removed them to save weight. If your convertible is missing its cocktail shakers, find some used ones and restore the exterior finish because as of this writing nobody seems to be reproducing them. Dealers selling used Camaro parts, such as Steve's Camaros and GM Sports Salvage, are the best sources for original used items like these.

Wheels

The 1967 and 1968 Z28s were the only models with wheels larger than 14 inches; they had 15x6-inch wheels. In 1969 Chevrolet went all out and upgraded the Z28s with 15x7-inch wheels. The most distinguishable factory Camaro wheel is the Rally wheel. In the late 1980s it became extremely popular to put Rally wheels on Camaros, and it became common for Camaro owners who didn't want the popular Cragar S/S five-spoke or "slotted mags" on their rides to install larger 15x8 Rally wheels from Corvettes in the rear and narrower 14- or 15-inch Rally wheels up front. A few companies offer custom-built Rally wheels if you're look-

The 1967 had the distinctive round markers in the grille. There were 34,411 SSs produced in 1967 (15 percent of total production). This is an ultra-rare Royal Plum SS 396/375 hp owned by David Boland and is one of few 1967 Camaros converted to a 427 for Yenko by Dick Harrell. The skinny BF Goodrich tires on 14x6 stock Rally wheels were no match for the power. (Photo courtesy Brian Henderson)

ing to install wider tires and keep the car looking somewhat factory.

The only other factory wheels offered on first-generation Camaros were the 1969 five-spoke stamped-steel SS wheel and generic stamped-steel wheels with many different styles of hub caps and trim rings, which are rarely seen these days. Most people have since removed them to upgrade to aftermarket wheels, or they flew off while driving.

We assume most owners want to install aftermarket wheels on their first-generation Camaro. The following short list contains sources for wheels that have a vintage feel—the design looks like a wheel that may have been available in the late 1960s or the design was available in 1967 through 1969. There weren't many aluminum wheels available back then, so we included a few extras for fun. Because this isn't a pro-touring book we left out billet wheels as well as more current designs not reminiscent of the 1960s feel.

The most recognizable factory steel wheels for first-generation Camaros are Rally wheels with the distinctive center caps and trim rings. This 1969 has the factory optional (PY4) Goodyear F70-14 Polyester Cord 4-ply tires. (Photo courtesy Tony Lucas Collection)

However, I have included wheels that are available in 15-inch diameter all the way up to 18-inch diameters because some owners will make modifications and upgrades for looks or performance. Keep in mind that a heavier wheel takes more horsepower to get up to speed than a lighter one. It also takes more braking power to slow a heavier wheel. Installing a heavy steel wheel will basically rob more performance off the line than a lighter aluminum wheel. Installing a heavy wheel also can overpower stock brakes, especially drum brakes. We never suggest installing extremely heavy wheels on manual drum brakes, or even power drum brakes for that matter.

Remember, the larger the diameter of the wheel, the heavier the wheel is going to be. Depending on construction and size, cast-aluminum wheels can be as heavy as, or heavier than, a stock steel wheel. Some of the wheels mentioned below have a cast center welded to an outer steel rim, which greatly increases the weight of the wheel. Check the weight of your wheels before making

170 HOW TO RESTORE YOUR CAMARO 1967–1969

MISCELLANEOUS MECHANICALS AND OTHER ITEMS

a purchase. If your car has extra horsepower and upgraded disc brakes, heavier and larger-diameter wheels are not as much of a concern for affecting performance.

Larger-diameter tires have the same effect as a heavy wheel, but 17- or 18-inch-diameter wheels usually wear lower-profile tires, which reduces the weight of the tire. Overall, the weight of a steel-belted radial is heavier than the original factory nylon-corded tires.

American Racing wheels offers a few vintage-looking designs that include the Torque Thrust line of wheels (Original A Spoke, D, and II), the 200S, and T70R.

American Racing offers a range of wheel sizes for each wheel design, but most are available from 15x5 to 15x8.5 inches (TTIIs and 200Ss up to 15x12 inches) as well as 16- and 17-inch-diameter wheels in popular widths.

Minilite was popular with the vintage Trans-Am racing back in the late 1960s and is still around today with Mag-style wheel sizes up to 15x10 inches and Sport wheels available in sizes up to 15x8 inches for Camaros.

Vintage Wheel Works has a version of the old Minilite (named V48) available in 16x8-inch and 17x7-inch to 17x11-inch.

VWW also has classic-designed 5-spoke wheels (V40 and V45) in popular widths in 15-, 16-, and 17-inch diameters.

E-T Wheels (aka Team III Wheels) has a version of the Minilite as well. It's named LT-III and comes in many popular widths in 15- and 17-inch diameters. Along with other cool nostalgic wheels, E-T Wheels offers a nice 5-spoke wheel in many widths in 15-, 16-, 17-, and 18-inch diameters.

Another mainstay on the market is Cragar, with the introduction of the famous Cragar S/S Super Sport 5-spoke wheel that was first introduced in 1964 and is still being produced. Multiple variants of the S/S Super Sport are available in chrome, aluminum, and painted aluminum and many widths of wheels ranging from 15- to 18-inch diameters which fit first-generation Camaros. In 2009, Cragar released a whole new set of one-piece aluminum wheels resembling its black G/T wheels and other classics from the 1960s.

Wheel Fitment

If you're buying wheels and tires that are larger in diameter or wider than stock and you're not going to have a shop test fit them, save yourself some grief by reading the following information ahead of time. You need to know about backspacing and have a good idea of what the physical limitations are.

Backspacing is extremely important to wheel fitment. It's the measurement from the rear lip of the wheel to the wheel's flat mounting surface on its back side. If you have a wheel from a late-model car it will have too much backspacing and will hit the suspension or inner body panels. If the wheel has too little backspacing, it will stick out too far and the tire will stick out of the fender edges and cause severe rubbing problems—at the very least, the outer fender lip will cut up the sidewall of the tire. If you are installing a stock wheel, you shouldn't have any interference problems.

Keep in mind that stock wheels from a car like a Corvette do not fit on your Camaro the same way. Every car has its own backspacing, even though the stock wheels appear to look the same doesn't mean they are the same.

The second important factor in purchasing wheels is the physical constraints of the usable area in the wheelwell. The most obvious limitation is the distance between the inner wheelwell panel and the outer lip on the fender. You can increase this usable space in the rear wheelwell by "rolling" the fender lip on the quarter panels. There are a couple of steel panels spot welded together that make the lip, so bending the lip upward and out of the way is not easy. That takes some time with a larger hammer whacking on the inside edge to curve the lip up and out of the way so the tire clears it.

Slow and steady gets the job done without damaging the outside of the quarter panel or the wheel-opening arch. Don't cut the lip off! If you do, you'll be cutting off the area of the lip that's welded together, weakening the panel and making wheelwell lip that much harder to roll upward. The new cut edge becomes like a razor to cut your tire if it ever comes into contact with it.

The front-fender lip is a whole different problem, because if you need to roll the lip in order to gain a little more clearance, you have to heavily trim the inner fenderwell panel where it bolts to the wheel-opening lip.

Keep in mind that the 1967 and 1968 cars have smaller wheelwells front and rear to run wider wheels than a 1969, which has approximately one additional inch of space in each wheelwell. The extra space on the 1969s helped racers fit wider tires and gave them a little bit more traction than in previous years. This was the biggest help in Trans-Am racing, where the rules stipulated that

CHAPTER 13

the fender opening could not appear altered for additional tire clearance. Some have speculated that Chevrolet made this change on the 1969 specifically to be more competitive in Trans-Am racing.

The other space constraint on all first-generation Camaros involves the front suspension. At minimum, too much backspacing causes interference with the steering arm and outer tie rod—and that's only the static attempt to mount the wheel. Then you have the space constraint of the necessity of steering the wheel from lock-to-lock, where a tire and/or wheel with too much backspacing contacts the sway bar and frame. A wheel with too little backspacing comes into contact with the fender and fenderwell.

Ride height is another limiting factor in wheel-and-tire fitment. If you're stuck in the 1970s or 1980s and still have air shocks installed, ride height may not be an issue and you can probably fit 15x10-inch Centerline Auto Drag wheels and some N50-15 Pro-Trac tires. If you're interested in reducing the nose-bleed stinkbug stance to a more ground-hugging ride height, be a little more conscious about your tire-and-wheel fitment. A lower car brings the body down around the tires where fender and tire contact starts to be an issue.

Not all tires are created equal and you have to take that into consideration when pushing the envelope for wider tires in the fenderwells. Even though the numerical tire sizes are the same for two different brands of tires, it does not mean they have the same physical dimensions. Some tires have a larger sidewall bulge or a wider tread contact patch than others, and either creates fitment issues.

The last factor is how square the car is. This means that factory tolerances of the rear axle being centered between the two quarter panels can be as much as 3/8 inch off center. Basically the rear axle may mount as much as 3/8 inch toward the right or the left quarter panel. This causes interference between the tire and the fender lip on one side and not the other when fitting larger tires in the rear. If there is more than 3/8-inch variance between the left and right rear tires, you've got more problems than a factory variance. This difference can be caused by any of several things: tired or bent leaf springs; rotten leaf-spring bushings; a broken leaf spring perch or alignment pin; a serious accident that caused body or frame damage, as well as a botched repair or a failure to straighten the frame or body after the accident.

The front subframe bolts to the body with six large bolts, which can shift when the body bushings have worn out. If your body bushings have been replaced, it's possible the person who replaced them didn't align the frame before tightening all the bolts. Front-end collisions and side impacts can also shift the subframe off center under the body.

Body bushings are not the only parts that rot and cause the subframe to shift. The subframe itself can also rust out and cause subframe alignment issues, which can only be corrected by repairing or replacing the subframe. Severe cases of the floorpan rusting out completely can also be a problem, but that would be a lot more obvious, since your seat would probably fall out or your feet would poke through the floorpan and you'd know something was very wrong.

There's a happy medium where you can fit a specific-size tire and wheel without a problem rubbing or hitting anything—1967s and 1968s will comfortably fit 15-, 16-, and 17x8-inch wheels on all four corners with 4¾ inches of backspacing. Of course, there can't be any serious problems with the car not being "square," the rear-axle assembly must be original to your car (not wider or narrower out of another car), your car has front disc brakes (drum and disc brake spacing is different), and your car isn't too low. The 17x8s fit with 235/45-17s in front and 255/45-17s in the rear without much, if any, rolling of the rear fender lip.

According to Rich Deans of Goodies Speed Shop in San Jose, California, "This 17-inch wheel-and-tire combo is the no-brainer fit for first gens, but you can fit bigger tires if you don't mind doing a little massaging to get them to fit."

The 1969s fit the same wheel and tire package very comfortably without issue, but it leaves more room in the fenderwells since the body is wider than the 1967 and 1968. The 1969 rear quarter with rolled lips can

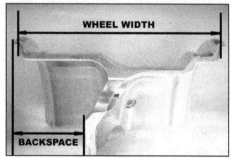

This wheel cutaway shows rim width and how backspace is measured on the back side of the wheel. This measurement determines how far the wheel mounts inward or outward inside the fenderwell. This measurement is critical when it comes to fitting wider wheels and tires on your car.

MISCELLANEOUS MECHANICALS AND OTHER ITEMS

First-generation Camaros look great with Vintage Wheel Works' nostalgic-looking wheels. Carl Casanova has 16x8 V45s with 4½ inches of backspace all the way around on his 1968 RS. Front tires were 245/50-16 and rears were 255/50-16 Some rear-quarter-panel lip-rolling was necessary. (Photo courtesy Carl Casanova)

safely swallow a 275 wide tire mounted on a 17x9 wheel with 4¾-inch backspacing. This 275 news is good for the drag-racing look with tires tucked in the quarters because racing 275/60-15 (L60-15) drag tires on 9-inch wheels fit too.

These 275s being on the larger side of the envelope may rub a little on the inner fenderwell when the suspension articulates while pulling in and out of driveways, but as long as your trunkpan pinch welds are not bent out, the inner fenderwells are flat and the tiny bit of rubbing is on a flat surface that won't cut the tire. Just make sure the entire surface is flat and no sharp edges on the surface or pinch welds are pointing toward the tire where it can cut the back side of the tire.

Every Camaro is slightly different due to brakes changing wheel position and factory variances. If you want to fit the largest tires possible in fenderwells of your specific car you should take your car to a custom tire and wheel shop like Goodies Speed Shop. It will test-fit a custom package using knowledge, measuring, and specialty fitment tools.

Tires

Correctly restored Camaros are not complete without the right tires. As far as tire performance goes, original reproduction tires do not ride or perform like new tires, so for safety reasons you may want to consider taking extra precautions when driving on them. There's a strong possibility you've been driving on radial tires for many years and you have grown used to the limitations of newer tire technology. Driving on older tire construction and technology makes the car perform differently during acceleration and braking, especially in panic-stop situations. Just be careful when driving your pride and joy.

First-generation Camaros came equipped with many different styles of tire from the factory. Depending on the year, model, or RPO chosen, your car could have come with complete blackwall, whitewall, white stripe (thinner stripe than the wider standard-option whitewall stripe), red stripe, or raised white letters.

If you're looking for a nice set of reproduction tires that would have originally come on your first-generation Camaro, check with Kelsey Tire for Goodyear tires. The 1968 and 1969 Camaro Z28s came with Goodyear E70-15 tires that had raised white letters that read "WIDE TREAD GT" along with the Goodyear name. These same tires are period correct for the 1969 Yenkos. Or check with Coker Tire. It makes reproduction Firestone and B.F. Goodrich Silvertown 660 tires for the first-generation Camaros.

For enthusiasts wanting a little more performance out of a tire or are upgrading to a 15-inch wheel that's wider than 7 inches or to a 16- or 17-inch" or larger-diameter wheel, you'll be forced to look at more than what Chevrolet offered as original standard or option tires. If you plan on doing any cornering or high-speed blasts, upgrade to some decent radial tires that match the driving application.

Tire technology has come a long way in 40-plus years for safety and performance. There are so many good tires out on the market to fit different budgets we can only offer up a few of the brands we've used with great success.

For regular street driving at a very decent price for larger-diameter wheels we've had good luck with Sumitomo tires and B.F. Goodrich Radial T/As for 15-inch tires. For more spirited driving around corners with larger-diameter wheels, we've found decent lateral traction predictability and streetability with BFG g-Force T/A KDWS and Yokohama Advan Sports. For excellent performance for the track and street we've had great luck safely driving an open-road-racing-prepped 1969 Camaro to a clocked 161 mph on BFG g-Force T/A KDW. We've had

CHAPTER 13

great success cornering on road-course tracks with the same tire.

Water Control

There are many ways to keep water from getting into your Camaro where it's not supposed to be: weatherstrips, gaskets and sealants around the doors, windows, trunk, engine compartment, accessories, and body seams.

Weatherstrips

When it comes to sealing the air and water out of the Camaro interior and trunk area, weatherstrip manufacturers offer products for two different applications: factory-correct look and fit, and aftermarket.

For factory-original-fit gaskets you can't go wrong if you can find original NOS gaskets still in their original packaging. Don't buy NOS weatherstrips in opened boxes; sellers may have slipped incorrect or ill-fitting newer gaskets in older boxes.

If you're restoring your Camaro to concours condition, be careful with reproduction weatherstrips. Over the years, we've had mixed results from SoffSeal (www.soffseal.com) and Metro Molded Products (www.metrommp.com). It seems that over the years these two companies' products have periodically gone through some changes and improvements.

The online enthusiast audience consensus says the Metro Supersoft and NOS weatherstrips are softer than other companies' products and are quite happy with the feel and fit of the Metro product. Metro weatherstrip construction is not hollow like some other weatherstrips, depending on application, and they are not multiple pieces bonded together. They are also a slightly different shape than other gaskets because Metro attempts to improve upon deficiencies in other weatherstrips.

Concours-restoration professionals prefer original NOS first, for obvious reasons. If they can't find what they need in NOS, they go to SoffSeal because these reproduction seals more closely mimic the factory weatherstrip look and construction.

If you install a weatherstrip that doesn't seem to seal properly right away, consider that the rubber may need a few days to conform to the new surfaces: the one it's attached to and the one it's supposed to seal against.

If a rubber weatherstrip or seal is rubbing on a window or another seal, lubricate the seal with some clear silicone paste, but keep in mind that silicone and uncured paint don't get along and will have a terrible reaction. We suggest installing weatherstrips on completely painted surfaces after new paint has been applied (or a good coat of paint is present). Be sure that the paint is completely cured before you attempt to install weatherstrips and seals.

Trunk Weatherstrip

Because the trunk weatherstrip is the only cut-to-fit weatherstrip on the first-generation Camaro, it seems to be the most feared seal to install. All the other seals are formed and only fit one way, unless you're really talented. Because the trunk weatherstrip draws the most inquiries, it deserves more detailed installation instructions.

The original trunk weatherstrips had a string in the center of the seal; later GM replacement seals had a hollow plastic tube in the center. These internal "back bones" performed a great job in keeping the gaskets taught and in place, especially around corners. Brian's restorations don't see much weather so in some cases he doesn't completely glue his in, so he can pull the seal out of the channel and clean it out periodically. Brian always uses original NOS GM trunk weatherstrips and does not suggest doing this with aftermarket reproduction seals.

Here's a cutaway view of two seals. The Metro Supersoft weatherstrip (left side) is slightly larger as well as a different shape and solid compared to the competing weatherstrip but, what you can't see is that the Metro seal is a much-softer, less-porous, and better-quality rubber.

The Metro trunk seal goes in easy and requires closing the trunk firmly and leaving it closed for a few days to get the seal set in. Trunk seals need to be cut to fit. The most common place for the trunk seal to end is along the back side by the latch assembly. Most end points are located to the left of the assembly but originals have been documented to end anywhere along the back side.

The trunk weatherstrip is the seal that you have to cut to fit. It's always longer than it needs to be. Just about every brand of weatherstrip is thick enough to make it hard to close the deck lid for the first few weeks until it gains a compressed memory. Install both of the rubber trunk bumpers before you start slamming the trunk lid to get it closed. Because the trunk seal is obviously going to have a spot where both ends meet, place the butted joint along the rear near the latch assembly. The seal is often located 10 inches to the left of the latch, but it can be located anywhere along the rear, and this has been documented.

Before gluing the weatherstrip into place, test fit it without any adhesive. Don't pull the weatherstrip tight when installing it, because as time passes, the seal settles in and doesn't fit as well.

When putting the seal into the channel, you can push a little extra weatherstrip in the channel. Try not to bunch too much into the channel or the gasket will compress and cause the gasket to get too firm and cause problems getting the trunk closed. On the other hand, don't put the seal in too short so you can easily get the deck lid closed. The rubber itself could just be inherently too thick for the application and needs time to compress properly.

The weatherstrip channel is about 135 inches long. Dispersed over the entire length of the seal, you could safely add 3 or 4 inches without having too much weatherstrip. Be sure to test fit your seal and close the deck lid. Cut the butt-joint ends very square with a sharp razor blade.

Before you glue your weatherstrip into place, make sure the channel is completely painted and the seams in the channel and at the quarter panel joints are sealed with a thin coat of seam sealer. This is necessary because if water gets past the seal, it will get into untreated areas and start to rust.

Follow these steps to install the seal:

Pull the seal out of the channel a couple of feet at a time.

Put a little adhesive into the channel and put the seal back.

Installing the seal all at once may allow the adhesive to dry too much.

Where the two ends butt together, they should join perfectly flat against each other. You can choose to leave them resting against each other or you can dab an extremely small amount of Super Glue around the butt-joint and stick them together. Don't get Super Glue on the sealing flap or sealing surface.

Don't close the trunk until the Super Glue and adhesive have cured, or you may glue your trunk shut.

If you need to tape your weatherstrip down to keep it in place or try to coerce it into a different shape, make sure you don't use the wrong tape. Only use Scotch 3M blue Painters Tape for Multi-Surfaces or Delicate Surfaces. If you use anything with stronger adhesive you can pull the top sealing surface off the weatherstrip, or worse. Don't leave the tape on the weatherstrip for more than a day or two or the adhesive may bite into the gasket.

Weatherstrip Adhesive

Chevrolet installed the first-generation Camaro weatherstrips with black adhesive. It wasn't very clean about it either. We've seen unrestored original cars with excessive adhesive on the surrounding painted surfaces. Nothing too bad, but enough to notice, so if you're going for a true factory-installed look it won't hurt if a lit-

We installed a new Metro weatherstrip with Kent Automotive weatherstrip adhesive and closed the windows against them for a couple of days to break them in. This method is correct for all weatherstrips to get them to "set" and get the correct shape.

tle adhesive oozes out of the channel. If you're looking for perfection where everything is as clean as a whistle and perfect, you're not performing a true "original" restoration. Keep in mind that the factory line workers had to install the seal in a matter of a few seconds. Getting it done cleanly was not the goal; getting it done was.

Use the proper weatherstrip adhesive for keeping gaskets in place. 3M Black Super Weatherstrip Adhesive (P/N 08008) or Kent Automotive Ultra-Stick (black or clear) are the best product to use for installing weatherstrips. The tube allows you to easily apply the adhesive in the weatherstrip channel. The adhesive remains flexible but keeps a strong bond.

Rubber Bumpers

The trunk lid has two rubber bumpers to keep the deck lid snug when it's closed and also to keep the deck lid from crashing down into the tail pan when slamming the trunk lid, which will chip your paint on the trailing edge. These rubber bumpers have been found in two different thicknesses, so pay attention to the fitment of the bumpers on your deck lid. If you leave too much of a gap, the bumpers will not be effective.

Source Guide

Ace Fuel Systems
975 Richard Ave.
Santa Clara, CA 95050
(408) 727-2838
www.acefuelsystems.com

Al Knoch Interiors
9010 North Desert Blvd.
Canutillo, TX 79835
(800) 880-8080
www.alknochinteriors.com

American Autowire Inc
150 Heller Pl. #17
Bellmawr, NJ 08031
(800) 482-WIRE
www.americanautowire.com

AMK Products Inc
(540) 662-7820
www.amkproducts.com

Arone Auto Body
136 West Elm St.
Homer City, PA 15748
(724) 479-3242
www.aroneautobody.com

ARP
1863 Eastman Ave.
Ventura, CA 93003
(800) 826-3045
www.arp-bolts.net

Baer Brakes Inc.
3108 W. Thomas Rd., Ste. 1201
Phoenix, AZ 85017-5306
(602) 233-1411
www.baer.com

Camaro Club
San Jose, CA
www.camaroslimited.com

CamaroPerformers.com

Camaros Limited Nor-Cal
Camaro Club
San Jose, CA
www.camaroslimited.com

Camaros Plus
5627 Kendall Ct.
Unit K
Arvada, CO 80002
(303) 420-6229

Camaros.net

Camaros.org

Campbell Auto Restoration
260 Cristich Ln., Unit A
Campbell, CA 95008
(408) 371-5522
www.campbellautorestoration.com

Carbuffs, Inc.
2281-A Via De Mercados
Concord, CA 94520
(925) 899-2648
www.carbuffs.com

The Carburetor Shop
1105 E. Walnut & Grand Ave.
Santa Ana, CA. 92701
(714) 556-2181
www.thecarbshop.com

Carter Fuel Systems
www.federalmogul.com

Champion Spark Plugs
www.federalmogul.com

Classic Industries
18460 Gothard St.
Huntington Beach, CA 92648
(800) 854-1280
www.classicindustries.com

Clevite77
www.mahleclevite.com

Coker Tire
(866) 513-2744
www.cokertire.com

D & R Classic Automotive
(888) Camaro-1 (888-226-2761)
www.drclassic.com

Damper Doctor
1055 Parkview Ave.
Redding, CA 96001
www.damperdoctor.com

Del City
2101 W. Camden Rd.
Milwaukee, WI 53209
(800) 654-4757
www.delcity.net

Detroit Speed
185 McKenzie Rd.
Mooresville, NC 28115
(704) 662-3272
www.detroitspeed.com

Dynacorn Classic Bodies
1400 Pacific Ave.
Oxnard, CA 93033
(805) 486-2612
www.dynacornclassicbodies.com

Eastwood
263 Shoemaker Rd.
Pottstown, PA 19464
(800) 345-1178
www.eastwood.com

E-T Wheels (AKA Team III Wheels)
1965 West 140
San Leandro, CA 94577
(510) 895-8880
www.etwheels.com

Federal Mogul
www.federalmogul.com

Fel-Pro Gaskets
www.federalmogul.com

Fine Lines Inc.
127 Hartman Rd.
Wadsworth, OH 44281
(800) 778-8237
www.finelinesinc.com

GM Sports Salvage
1964 Old Oakland Rd.
San Jose, CA
(408) 432-8498
www.gmsportssalvage.com

Goodies Speed Shop
345 Lincoln Ave.
San Jose, CA 95125
(408) 295-0930
www.goodies-speedshop.com

Goodson Tools
156 Galewski Dr.
PO Box 847
Winona, MN 55987
(800) 533-8010
www.goodson.com

Griffin Thermal Products
100 Hurricane Creek Rd.
Piedmont, SC 29673
(800) 722-3723

Gromm Racing Heads,
Balancing & Engine Machine
664-J Stockton Ave.
San Jose, CA 95126
(408) 287-1301

Heartbeat City
15081 Commercial Dr.
Shelby Twp., MI 48315
(586) 226-8811
www.heartbeatcity.net

Hotchkis Performance
12035 Burke St., Suite 13
Sante Fe Springs, CA 90670
(888) 735-6425
www.hotchkis.net

Jim Dyer Chevy Classics
(209) 941-2112
www.jimdyerchevyclassics.com/

Joe Gibbs Racing
13201 Reese Blvd. W.
Huntersville, NC 28078
(866) 611-1820
www.joegibbsdriven.com

Just Dashes
5941 Lemona Ave.
Van Nuys CA 91411
(818) 780-9005
www.justdashes.com

KBS Coatings
2502 Beech St. Suite 100
Valparaiso, IN 46383
(888) 531-4527
www.kbs-coatings.com

Keisler Automotive Engineering
2250 Stock Creek Blvd.
Rockford, TN 37853
(865) 609-8187
www.keislerauto.com

Kelsey Tire
(800) 325-0091
www.kelseytire.com

Kent Automotive Products
(888) 937-5368
www.kent-automotive.com

Lateral-G.net

Legendary Auto Interiors, Ltd.
121 West Shore Blvd.
Newark, NY 14513
(800) 363-8804
www.legendaryautointeriors.com

Lucas Restorations
(443) 212-5580
www.lucas-restorations.com

Mahle
www.mahle-aftermarket.com

Metro Moulded Parts
11610 Jay St. N.W., Box 48130
Minneapolis, MN 55448
(800) 878-2237
www.metrommp.com

Milodon
2250 Agate Ct.
Simi Valley, CA 93065
(805) 577-5950
www.milodon.com

Moog
www.federalmogul.com

Morris Classic Concepts
(864) 987-0032
www.morrisclassicconcepts.com

Muscle Car Guys
7100 East Meadow Ln.
North Little Rock, AR 72118
(501) 812-5993
www.themusclecarguys.net

Muscle Car and Corvette Nationals
Bob Ashton
P.O. Box 182068
Shelby Township, MI 48318-2068
(586) 549-5291
www.mcacn.com

OER
18460 Gothard St.
Huntington Beach, CA 92648
(800) 955-1511
www.oerparts.com

Optima Batteries
Johnson Controls
(800) 867-8462
www.optimabatteries.com

Original Car Radios
www.originalcarradios.com

Paragon Corvette Reproductions
8040 South Jennings Rd.
Swartz Creek, MI 48473
(800) 882-4688
www.corvette-paragon.com

Performance Stainless Steel
2851 Research Park Dr. Unit A
Soquel, CA 95073-2093
(831) 713-5520
www.performancesst.com

Power Probe
890 Mariner St.
Brea, CA 92821
(800) 655-3585
www.powerprobe.com

Prothane
3560 Cadillac Ave
Costa Mesa, CA 92626
(714) 979-4990
www.prothane.com

Pro-Touring.com

Quality Restorations
13983 Humo Dr.
Poway, CA 92064
(858) 271-7374
www.qualityrestorations.com

RaceHome.com
PO Box 8232
San Jose, CA 95155-8232
www.racehome.com

Retro Sound
www.retrosoundusa.com

Screamin' Performance
631 Gladiola St.
Merritt Island, FL 32952
(312) 452-5996
www.screaminperformance.net

Soff Seal
104 May Dr.
Harrison, OH 45030
(800) 426-0902
www.soffseal.com

Speed Pro
www.federalmogul.com

Steve's Camaros
1197 San Mateo Ave.
San Bruno, CA 94066
(650) 873-1890
www.stevescamaroparts.com

Super Car Workshop
(724) 532-1975

Thermo-Tec Automotive Products
P.O. Box 96
Greenwich, OH 44837
(800) 274-8437
www.thermotec.com

Trim Parts
2175 Deerfield Rd.
Lebanon, OH 45036
(513) 934-0815
www.trimparts.com

Victor Reinz
www.victorreinz.com

Vintage Air
18865 Goll St.
San Antonio, TX 78266
(800) 862-6658
www.vintageair.com

Vintage Musclecar Parts
Eric Jackson
306 Lang Ct.
Union, OH 45322
(937) 836-5927
www.vintagemusclecarparts.com

Year One
P.O. Box 521
Braselton, GA 30517
(800) Year-One (800) 932-7663
www.yearone.com

Yenko.net

CPSIA information can be obtained
at www.ICGtesting.com
Printed in the USA
BVHW012313220620
R10889200001B/R108892PG581954BVX1B/1